It's Really About Time

It's Really About Time

The Science of Time Travel

John Oliver Ryan

Tahilla Press
Woodside, California

Printed in the United States of America.

ISBN: 978-1-7342643-0-2 (softcover)
ISBN: 978-1-7342643-1-9 (ebook)

Please direct comments or criticisms to:
john@tahillapress.com

Address permission requests to:
info@tahillapress.com

For information on purchasing multiple copies of this book at reduced prices, please inquire at info@tahillapress.com

Tahilla Press, Woodside, California

ACKNOWLEDGEMENTS

I wish to thank those of my friends and family who took the time to read and comment on drafts of this book. Their corrections and suggestions for improvements and clarifications are much appreciated.

They are:

Friends: Gibson Anderson, Mark Belinsky, Don Bogue, Francois Cazenave, Helmut Dorn, Dr. Roger Downer, Paul Fehrenbach, Casey Handmer, Dr. Josh Hogan, Jim Holzgrafe, Carol Holzgrafe, Bill Krepick, Vitaly Lazorin, Richard Mattern, Niall Norton, Val O'Connor, Norma Piras, Sergio Piras, Pat Shiel, Claudio Todeschini, Steve Weinstein, and Peter Wonfor.

Family: Chris, Fiona, Katrina, Pat, Pauline, Stephen, and Torrey.

Any remaining errors and inconsistencies are entirely attributable to me, aided and abetted by my muse, the estimable C. C. Sauvignon.

Book design and illustrations by Jim Holzgrafe.

It is not widely appreciated that the possibility of traveling forward in time – by months, years or even centuries – far from being science fiction, is well within the realm of established science.

Contents

APPENDICES

AUTHOR'S NOTE

Before Albert Einstein made his astonishing debut on the world's scientific stage in 1905, our understanding of the physical universe rested firmly on the three laws of motion and the law of gravity discovered by the great English scientist and polymath Sir Isaac Newton, in the late 16th century. These laws allowed scientists at the time – and still allow them today – to predict, with exquisite accuracy, a wide range of phenomena from the motions of the planets to the behavior of tides and of objects moving under the influence of gravity here on Earth. But in the latter part of the 19th century severe difficulties arose when applying these laws to explain the behavior of light; it's not much of an exaggeration to say that scientists at the time were in a deep philosophical funk from their decades-long struggle to understand the matter.

Enter twenty-six-year-old Einstein; guided largely by his own aesthetic ideals on how the universe ought to be ordered, he conjured up his Special Theory of Relativity which completely and elegantly cleared up those difficulties, and in so doing fundamentally altered mankind's conception of our universe.

Einstein introduced Special Relativity in 1905, yet over a century later its teachings remain strange – weird, even – to many people, including scientists from other disciplines. Surprisingly – and unlike many ideas in physics – those teachings are easy enough to state in plain English without a boatload of attendant mathematical equations. Yet they are so at odds with our everyday experiences that they can be more difficult to believe than they are to understand.

Of the several extraordinary predictions of Special Relativity, none was as jarring to my own worldview as the possibility of traveling to the future – to a time that could be years or even centuries hence.

Another of its predictions was embodied in the now iconic $\mathbf{E = mc^2}$ equation, which exposed and quantified an equivalence between energy and matter. As unanticipated as that discovery was, in my opinion it's not quite as mind-blowing as the possibility of time travel – though close-up witnesses to a nuclear explosion might beg to differ.

The very idea of time travel intrigued and delighted me when I first encountered Special Relativity as a seventeen-year-old physics student. The textbooks I had access to were of little help as they focused almost exclusively on the underlying mathematics, their main purpose being to teach students just what they needed to know in order to pass their physics exams. Thus, I learned the theoretical underpinnings of the subject without having a good conceptual understanding of how and why such an extraordinary thing as travel to a future time could be possible. Later in life I got around to revisiting the subject with a view to a deeper understanding. Having achieved that – at least to my own satisfaction – I decided to put it in writing.

This book was written for intellectually curious people – with or without a science or mathematics background – who have heard about time travel and either dismissed it as pure science fiction or were intimidated by its "difficult to understand" reputation.

Many popular books on relativity when they address time travel often lose their readers in long discussions of the so-called Twin Paradox, from which they seldom fully extricate themselves. The reader is frequently left muttering "but, but…". My hope is that when you finish reading this you will feel confident that you could explain the scientific basis of time travel to your friends. But please do so only if asked; good friends are scarce…

My goal with this book is to present readers with a self-contained, full and coherent explanation of why time travel is possible (in principle if not yet in practice) with enough historical embellishment to fill out the story and put it in context. To this end, the earlier chapters include an introduction to some basic ideas in physics that are quite straight-forward and easy enough for all readers to follow and understand. The only prerequisites are that readers have a clear grasp of everyday notions of distance, time , and speed and are not intimidated by very large or very small numbers. Speed is simply distance traveled in a given time and is usually expressed in feet (or meters) per second, or in miles (or kilometers) per hour – but could equally well be expressed in furlongs per fortnight by a mischievous science teacher.[1]

I make much use of thought experiments. These are techniques for analyzing problems from the comfort of one's armchair, with feet up and a glass of favorite tipple to hand. However, for the results of thought experiments to be valid they must not violate established laws of physics, though they are allowed to invoke experimental setups that may be beyond our present abilities to implement or that are downright dangerous to enact.

Some of these thought experiments take place on Earth and others in outer space. There's nothing special about outer space as far as relativity goes; it's just that when a thought experiment involves objects traveling at tens of thousands of kilometers *per second*, we need lots of (ahem) space to do it in or we'd be constantly bumping into other objects.

A few of these thought experiments involve measuring the speed of a bullet or a sound wave or a light beam. For readers who'd like a basic understanding of how such measurements can be made, see appendix 1 (Measuring Speed).

You will find very few mathematical equations in this book, and you will not need to comprehend them in order to follow the discussion. It's a remarkable fact that as strange as the notion of time travel can be to get one's mind around, the famous Lorentz equation that underpins it is about as simple as physics equations get. It can be derived using little more than the Pythagorean theorem (*the square on the hypotenuse is equal to the sum of ... yada, yada, yada.*) which most of us were force-fed in geometry class in our early teens. But to guard against a panic attack in readers with refined artistic sensibilities, the derivation of that equation – little bitty piece of algebra though it be – has been banished to appendix 5 (Proof of the Lorentz Formula).

The speed of light plays a starring role in this story since Special Relativity was devised primarily to explain the strange results obtained when measuring light's speed in different situations. Since the speed of light is almost exactly 300,000 kilometers per second (km/sec) in the

---

1. By pure coincidence, one furlong per fortnight is almost exactly equal to one centimeter per minute; that's a snail's pace even by snail standards.

metric system versus about 186,282 miles per second in the imperial system, in the interests of using round numbers and consistency we will use the metric system throughout.

Light-years, light-minutes, and light-seconds are each measures of distance, not of time. A light-year is the *distance* light travels in one year – about 10 trillion (i.e., 10 million million) kilometers. The sun is about 150 million kilometers away, which is equivalent to about eight light-minutes; i.e., light from the sun's surface takes eight minutes to reach Earth.

You will encounter very small units of time since light travels 300 meters in one microsecond, or 30 centimeters (about a foot) in one nanosecond. We are all well used to distances like these in our daily lives, but not to such small time scales, unless one happens to be an electronics engineer or nuclear physicist. A microsecond is a millionth of a second and a nanosecond is a billionth of a second. Modern electronic instruments are capable of accurately measuring the durations of events lasting far less than a nanosecond; indeed, the average personal computer sold in 2019 is capable of carrying out a basic instruction (e.g., adding two numbers) in a nanosecond or so. But to a nuclear physicist, a nanosecond is a lifetime; a particle of light (called a photon) can traverse an atomic nucleus in about a trillionth of a nanosecond.

You may encounter some terms that are new to you, such as electric field, magnetic field, electromagnetic wave, etc. If not defined in the text, you won't need to understand them in any greater depth in order to follow the discussion. They are included mainly because they are integral to the historical development of Special Relativity, but also because they provide readers wishing to dig deeper with key words to enter in on-line searches.

There is one concept that we will use quite a lot that may be unfamiliar. That's the concept of a reference frame, or just a "frame." You will come across situations in which person A, who considers herself to be stationary, is observing person B who is moving relative to A. Person A, who might be standing on a railway platform, is said to be in the stationary or reference frame and person B, who might be riding in a passing train, is said to be in the moving frame. We will learn later when we consider the Principle of Relativity that there is no such thing as absolute motion and that either A or B may consider herself to be stationary and the other to be moving, depending on the details of the thought experiment we are describing.

Of necessity, a book of this kind for its intended audience can only skim the surface of Special Relativity. A deep dive into the subject would likely be a multi-year endeavor even for someone already well-versed in general physics and possessing good mathematical abilities. Nevertheless, in choosing the various topics I felt were essential to providing a seamless explanation of how time travel is possible, I took care to present them accurately, even if, to paraphrase Emperor Joseph's complaint about The Marriage of Figaro *in* Amadeus, *I used "too many words."*

Introduction

IN WHICH WE PROVIDE SOME HISTORICAL CONTEXT, ALONG WITH AN OUTLINE OF HOW THIS BOOK IS STRUCTURED

FOREWORD

In April 1906 the city of San Francisco was shaken to its core by a massive earthquake, its streets ripped open and its buildings toppled; the world reacted with appropriate shock and horror.

Few knew that a few months earlier, in September 1905, Albert Einstein's Special Theory of Relativity had been published. Groundbreaking to say the least, it toppled some of the long-standing edifices of physics! And while the Great Earthquake has faded to a distant memory, Special Relativity now forms the bedrock of much of modern science and Einstein is celebrated as one of the greatest scientists of his own or any other era.

The scientific basis for travel to a future time rests firmly on one of that theory's predictions, namely that the rate at which time flows is *not* the same everywhere and for every person – contrary to what the direct evidence of our senses suggests – but is in fact different for people or objects in motion relative to each other. In the intervening years, numerous experiments have verified that prediction. Moreover, many modern technologies such as the Global Positioning System (GPS, also known as SatNav) and particle accelerators, such as the Large Hadron Collider near Geneva and the Stanford Linear Accelerator near San Francisco, *must* account for this variability of the passage of time in their design details. Failure to account for it could cause kilometer-sized location errors in the case of GPS and would render particle accelerators essentially useless.

Time travelers of science fiction lore are usually seated in a chamber with walls full of dials and meters and flashing lights doing Lord-knows-what and leading, after a massive build-up of sound and light, to the instantaneous transportation of the travelers to a different time and place.

But really, all you have to do, as we'll see, is travel at extremely high speed for a period of time, and whether you intended it or not you will have travelled some amount of time into the future of all those who stayed put. And the reasons we're not all doing this for fun or adventure are two-fold: the speeds required – around a million times that of jetliners – are a bit beyond our present technologies, and it would strictly be a one-way journey.

No round trip tickets available – at any price!

FIRST, A LITTLE HISTORY

Prior to Einstein, the assumption that the flow of time was the same everywhere throughout the universe was almost an article of faith for scientists, philosophers, and just about anybody who had given the matter some thought. The afore-mentioned 17th century scientist Isaac Newton had all but etched it in stone and delivered it as a commandment when he declared:

> *"Absolute true and mathematical time, of itself, and from its own nature flows equably without regard to anything external."*

Newton had an equally strong opinion on the nature of space[2]:

> *"Absolute space, in its own nature, without regard to anything external, remains always similar and immovable."*

Up until the end of the 19th century, Newton's ideas about time and space underpinned all known scientific laws – laws which were hugely successful in explaining a wide variety of natural phenomena – so scientists of the time had every reason to expect that those laws would still hold true in whatever new domain they might be tested.

The launch pad for Special Relativity was the growing conviction toward the end of the 19th century – inspired by both empirical facts and theoretical considerations – that the speed of light as measured by any and all observers was independent of the speed of the source of the light and also independent of the speed of devices making the measurements.

To get a feel for how strange this must have seemed at the time, imagine a flashlight mounted at the front of a spaceship, arranged to flash on very briefly once every second. Imagine also that there's a second spaceship some distance ahead and equipped with a device that can measure the speed of these light flashes as they pass by it. *The speed of these light flashes as measured from the second spaceship would always*

2. Space, as in the 3-dimensional space which encompasses every object in the universe.

be the same 300,000 km/sec, irrespective of whether the two spaceships were stationary or were moving toward or away from each other, and no matter how fast. Even if the second spaceship were moving away from the first at half the speed of light, it would still measure the speed of the incoming flashes at 300,000 km/sec.

This *constancy of the speed of light* was utterly inexplicable at the time; *it defied common sense.* For it was well known back then that the measured speed of a material object (e.g., a bullet) relative to an observer depended on both the speed (if any) of the rifle that fired the bullet and on the speed (if any) of the observer making the measurement. It was also well known that the speeds of various kinds of waves were constant relative to the medium in which they were carried. Thus, for sound waves in air, water waves in a lake, or earthquake waves in the ground, each had a characteristic speed that depended on the physical properties of the medium – properties such as density, viscosity, etc. For example, sound waves travel through air at sea level at about 340 meters per second; this speed is relative to air itself.

However, the *measured* speed of each kind of wave also depends on the speed of the measuring apparatus *relative to the medium carrying the wave.* So, the speed of a sound wave as it passes by you and your measuring apparatus depends on whether you are stationary or moving through the air toward or away from the source of the sound. And even if you were stationary, its speed would be affected if the air carrying the sound were moving, i.e., if there were a wind blowing.

So, whether light consisted of microscopic material particles or was a wave of some sort – and the answer to that was not certain at the time – the fact that its measured speed appeared to be constant for all observers whatever their state of motion could not be explained; it was a huge and glaring affront to Newton's well established and successful laws. Light clearly behaved differently from both material particles and various types of waves, and all attempts to account for those differences with minor adjustments to the physics of the day had failed. As the decades passed and the problem remained unresolved, it became clear that something was amiss at the very foundations of physics. Since the problem had to do with the speed of light, and since speed is simply distance (i.e., space) traversed in a given time, it became clear that there had to be something wrong with the then-prevailing conceptions of space and time as enunciated by Newton.

When we say that certain things were well known at the time, it's not just that scientists of that era could predict with great accuracy, for example, the speed of a bullet fired from a rifle having a certain muzzle velocity when the rifle itself is also moving at a certain speed – as measured by an observer moving at a different speed; more to the point, the equations underpinning such predictions flowed logically from Newton's assumptions of a) the complete independence of space and time, and b) that how we experience space and time are the same everywhere and under all conditions.

But Newton was wrong about that...

TRAVELING TO THE FUTURE

The science underpinning travel to the future is the main topic of this book, yet the very idea of it may seem absurd; for how can one travel to a future time or era that, by definition, hasn't happened yet? Here's the general idea:

Imagine that there's a magical place in which time passes at a mere one-tenth the rate to which you're accustomed. Let's call it a Time Warp Environment (TWE). If you were to enter this TWE, here's what would seem different to you: *Precisely nothing!*

Everything inside this TWE would appear completely normal. Your hair and nails would grow at the same rate they always did and you would age at the same rate you always did *according to your wristwatch and all the clocks within that TWE.* Most importantly, your sense of the passage of time would be exactly the same as it was in the world you were used to; i.e. the passage of an hour according to the clocks within the TWE would still seem like an hour to you, despite the fact that those clocks – *and therefore time itself* – were running ten times slower than normal.[3]

Let's say you enter this TWE on January 1st, 2100 and remain there for one year according to both your own watch and the clocks within this TWE. When you exit to your original world again, *your* watch would of course read January 1st, 2101, but the clocks in your original world would read January 1st, 2110.

3. This follows directly from a consideration of the fundamental physics and chemistry of our brains. Even though medical science is a long way from a detailed understanding of brain function, there is every reason to believe, based on present knowledge, that at its most basic level, brain function is mediated by physics and chemistry, rather than by some as-yet-undiscovered mysterious forces. That being so, since all physical and chemical processes within the brain will also be slowed by a factor of ten within our hypothetical TWE, so will the particular brain process, whatever it may be, that gives us our sense of the passage of time. See appendix 11, Question 2 ("Q2") for a fuller discussion of this.

So, by entering this TWE and remaining there for one year, when you return to the world you left you will have effectively travelled ten years into *its* future. If you remain inside this TWE for *ten* years by your watch, when you return to the world you left you will have travelled a *hundred* years into its future.

The foregoing is what we mean by traveling to the future - in the scientific rather than the science-fiction sense; you enter an environment where time passes more slowly and remain there for some period while time continues to pass at the normal rate in the world you left. So, when you return to that world, it will necessarily be to a future time there. Had you left an identical twin behind when you first entered the TWE, on returning you would find your twin to be older than you in all respects and perhaps even long dead.

Note, however, that it will not be to *your own future* that you will have travelled; you will not emerge in a future time and discover, or suddenly become, an older version of yourself. Rather, you, at the age and physiological state you were in when you left the TWE, will simply emerge *as is* to a future time in your original world, a world which had continued on its merry way without you while you were aging more slowly in your TWE. And if you had left an identical twin behind, on returning that twin would see you emerge *as a younger twin*, if he or she were still alive.

Indeed, how could it be otherwise? If it were possible to travel to *your own future*, you could in principle visit your grave and end up wondering what or who lies therein! A serious logical and philosophical problem to say the least... To accomplish such a feat, you would have to produce a perfect clone (You 2)[4] of yourself (You 1) that remains behind to live its life while You 1 enters a TWE for a trip to the future, as outlined above. On arriving at a future time You 1 could meet You 2 at a pre-agreed location – or even visit his grave... Now *that's* science fiction!

Well, you may say, the foregoing is all fine and dandy as a matter of idle speculation, but where can I find one of those Time Warp Environments?

4. *Pace*, Bono

In a limited sense, they're all around us…

We will learn later, in detail, that one of the consequences of Einstein's Special Theory of Relativity is that time passes more slowly in a moving vehicle (be it in a car, train, spaceship, etc.) relative to an observer *deemed* to be stationary. This goes by the somewhat intimidating name of *relativistic time dilation*, or just *time dilation*. But no need to be intimidated; the concept is easy to understand (though it may be harder to believe).

Here's the basic idea: Stationary observers see (*infer*, more correctly) clocks (and all other time-dependent processes) that are located in moving vehicles to be running slower than their own clocks. Later we'll learn why this is so, and by how much. For now, we'll just note that at the speeds we are used to in our daily lives, the amount of slowing down of time is completely and utterly negligible. Even at the speed of a jetliner, it amounts to only about one billionth of a second (a nanosecond) per hour; two-tenths of nothing, one might say. But as such vehicles approach the speed of light, the amount of slowing down becomes very obvious. For example, events that take one hour on Earth, according to Earth-based clocks, will take two hours in a spaceship traveling at about ⅞ of light speed, as viewed or inferred by an Earth-based observer using an Earth-based clock, though just one hour by the spaceship's passengers looking at their own internal clocks.

We say that time has slowed down in the spaceship; it has become expanded (i.e., *dilated*) relative to the outside observer. Hence the term *time dilation*.

So in a limited sense, TWEs are indeed all around us, in the form of vehicles traveling at any speed. But in order to be able to slow the passage of time by a meaningful amount those vehicles would have to travel at enormous, presently unattainable speeds.

Jumping ahead a bit, to actually travel to a future time you would enter a spaceship, accelerate up to a sizeable fraction of the speed of light and continue at this speed for a period of time. When you eventually slow down and stop, whether it be on some distant planet or right back where you started, you would have travelled to a future time. And how far forward in time you would have traveled will depend on the speed of your spaceship and the amount of time you spent traveling. By way of example, if you travel for one year, as measured by your watch, at 99.5% of light speed, at the conclusion of your trip you will have moved ten years into the future, regardless of where you chose to end up, versus just one year had you stayed home.

Traveling forward in time by moving at extremely high speeds for extended periods will be the main mechanism for time travel we will focus on in this book. But there is another mechanism, also discovered by Einstein, that has its own collection of Time Warp Environments, which we'll now describe briefly.

In 1916, eleven years after completing his Special Theory, Einstein published what came to be known as his General Theory of Relativity, which expanded on his Special Theory to encompass gravity and objects undergoing acceleration. A remarkable aspect of this theory was its prediction that the passage of time would be slower in the gravitational field of a planet or star than in the virtually zero gravity environment of intergalactic space, and the stronger the gravity field, the more that time would slow down. This effect also influenced the design of the GPS system, as we'll see later.

And that's all we'll need to know about gravity-induced time dilation for our present purposes.

Poor Newton! In the course of just ten years Einstein first upended Newton's conceptions of time and space with his Special Theory of Relativity and then went on to upend his ideas on gravity with his General Theory of Relativity. But Einstein, being the gracious person that he was, and fully appreciating the awesome contributions Newton had made to physics and mathematics, wrote in his Autobiographical Notes: *"Newton, forgive me. You found the only way which, in your age, was just about possible for a man of highest thought and creative power"*.[5]

5. Many books and articles that discuss Einstein's and Newton's physics tend to be overly dramatic when describing how *Einstein completely overthrew Newton's physics*. The fact is that except in areas of extreme physics (colossal speeds or extremely strong gravity) as discussed in appendix 7, Newton's physics provides extraordinarily accurate descriptions of reality. Indeed, a deep working knowledge of it remains essential for physicists, most engineering disciplines,

Before moving on, it's essential to recognize that time dilation isn't about the notion that the mechanisms of clocks, whether mechanical or electronic, might be disturbed and rendered inaccurate by high speed motion. Time dilation causes an equivalent reduction in the number of beats of a moving person's heart per stationary observer's minute as it does the number of ticks of a moving clock per stationary observer's minute. Perhaps the simplest way to visualize time dilation is to imagine you're watching a movie running at a slower than normal rate. *Every* time-dependent event in the movie will be observed to be running slower by the same factor, whether it's the movement of the hands on a watch, the swinging of the pendulum on an old grandfather clock, the rate at which the sand flows through the orifice in an hour-glass, or the beating of someone's heart. Time dilation is about differences in the apparent *rate* at which time flows. Clocks, watches, hour-glasses, etc. are just devices humankind has devised to keep track of that rate.

---

and a host of other professions. All the more remarkable given that Newton did most of his work over 300 years ago. He was born on Christmas day in 1642; a gift to the world that keeps on giving.

OUTLINE OF THIS BOOK

This chapter provides a heads-up on what's to follow and is intended as a reference for readers as they progress through the remainder of the book.

We begin with a love story that should help readers gain a clearer understanding of what it means to be able to travel to a future time. Though its details are obviously contrived (it *is* fiction), all aspects of the story are technically correct. All that's missing in order to enable it in practice is a really high-speed spaceship and willing participants!

Next we expound a little more on time travel, using a few limericks for light relief.

Having set the stage in the Introduction, Section I introduces some basic concepts in physics that should help readers more fully appreciate the problem that Einstein solved in 1905. This section should be easy to follow, even for readers who assiduously conspired to avoid any traces of science, technology or mathematics in their education; though perhaps not so easy that one can follow it with one eye and half of one's brain focused on a favorite TV program.

We start our physics primer by introducing the idea that there is no such thing as absolute motion and that instead, all motion must be considered as being relative to some place or thing *that we arbitrarily deem to be stationary*.

Next we introduce the Principle of Relativity, one of the most important concepts in all of physics and which dates back at least to Galileo's time in Renaissance Italy. This principle states that all laws of physics are the same in all frames (places or vehicles, if you prefer) that are moving at constant speed relative to each other.

A word of warning: The Principle of Relativity played a major role in Einstein's development of his Special Theory of Relativity and we will refer to it many times in this book. *Take care not to confuse the Principle of Relativity with the Special Theory of Relativity!*

Next we take a close look at how material objects and sound waves travel, and how their measured speed is affected by the speed of their source and the speed of the person or instrument performing the measurement. We discuss how it seemed entirely reasonable (before Einstein) to assume that the measured speed of light would also depend on the speed of the source of the light and/or on the speed of the observing instrument, in a manner that depended on whether light was a particle or a wave.

We show via a number of illustrations precisely what we mean when we say that the measured speed of light is completely unaffected by the speed of its source or the speed of the person or instrument carrying out the measurement, and how this fact – inexplicable at the time – caused a crisis in physics toward the end of the 19th century.

In Section II we describe how Einstein dramatically resolved this crisis, by daring to imagine a world in which time and space are not separate entities but instead are deeply intertwined. Moreover, his solution – the Special Theory of Relativity – ended up impacting the scientific world in ways that were at the time unimaginable, and even today can be a source of awe and wonder for newcomers to the subject.

Next we take a close up look at the difference between Newton's conception of time and the new conception introduced by Einstein.

Then we introduce a physics concept called the Doppler effect which was known and well understood many decades before Einstein came on the scene. It's an important concept to understand because it can cause an *illusion* of time dilation and its opposite, time contraction, and understanding it will enhance our understanding of what true time dilation means.

Lastly in our physics primer, we introduce another thought experiment that explains how an object traveling initially, say, from west to east, will move after it has received a push in the north-south direction. This will help us follow a later thought experiment.

At this point we will have covered all the basic physics needed in order to follow Einstein's proof that time passes slower in moving frames than in frames deemed to be at rest.

We ease up a bit with an Intermission, in which we review what we've learned already and set the stage for our assault on the central question: Why does time slow down in moving frames, and by how much, and how can this fact allow for travel to a future time?

Section III opens with our most important thought experiment, by which we prove that time really does slow down in moving frames relative to stationary observers, and we state the formula by which we can calculate just how much time slows down for a given speed. We also learn by revisiting the Principle of Relativity that this observed slowing down of time works both ways. Since either party may consider herself to be stationary and the other to be moving, it follows that each will see the other's time to be running slower relative to her own, and we show that this is not the logical contradiction it may appear to be at first sight.

In Section IV we carefully follow the high-speed space journey of Harry from our love story and prove conclusively that he really would have aged less than Carrie, thereby accomplishing the main purpose of the book.

In the Appendices we offer explanations on how various measurements of speed can be made and we prove some points that we glossed over in the main text. We also discuss some general aspects of scientific inquiry and show that an understanding of how the passage of time is affected both by speed and by gravity was critical to the design of the Global Positioning System.

STORY TIME

Sometime in the far future, 18-year-old Carrie, a high-school student, and 26-year-old Harry, a post-doc in the physics department of a nearby university, have fallen in love. They plan to marry on Carrie's 21st birthday, but their plans are thwarted when Carrie's grandmother passes away and in her will leaves Carrie her beautiful house overlooking San Francisco Bay.

But there were strings attached...

You see, Grandma had been a world-renowned cardiac surgeon and had been thrilled to learn a few months before her death that Carrie had plans to follow in her footsteps. But she knew that Carrie had no real appreciation of the study regimen and the long workdays she would have to endure in order to complete her university and residency requirements before she could even begin to practice her chosen profession. From her own experience, Grandma appreciated the stress that this would place on Carrie's marriage and she was concerned that the age difference between Carrie and Harry would just add to that stress.

So she hatched a plan that she was sure would be in Carrie's best interests. The terms of her will stipulated that Carrie would only receive the house if she did not marry before her twenty-eighth birthday and had also completed her studies. And further, Carrie must marry someone close in age to her own physiological age!

The will was legally ironclad; if Carrie wants to own that house some day, she will have to abide by its conditions. She is devastated; she and Harry are not well off and it would be years before their combined salaries would allow them to afford a gorgeous home like that, if ever. But worse still, the requirement that she marry someone close to her own age would prevent her from ever marrying Harry.

Over the following weeks, Carrie and Harry talk it over and finally come to a decision; they conclude that they love each other so much that they will reject Grandma's legacy and continue with their previous life-plans.

Harry, though, has an unusually creative and logical mind and doesn't give up easily. He reads and re-reads the terms of the legacy, looking for a loophole. On his seventh reading he freezes at the phrase "her own *physiological* age."

Why did Grandma add that adjective? he wonders. *What does the term 'physiological age' even mean? Why not just 'age'? Aren't they the same thing?*

Since it's the only uncertain phrase in the entire document – all the others being clear and unambiguous to a fault – Harry turns it over and over in his mind trying to make sense of it. Harry knows that Grandma had written that will herself, every word of it, each one carefully chosen to faithfully convey her wishes. So he focuses hard on that seemingly redundant adjective: *physiological.*

Physiological. How could one's physiological age be different from one's regular age? If you're twenty years old according to your birth certificate, then, all else being equal you will have aged physiologically by twenty years, no more, no less.

Harry's mind goes into overdrive...

What if all else were not equal? How could I make "all else" not be equal? How could a 26-year-old person be physiologically like an 18-year-old person? Impossible! No way to wave a magic wand and make me look and feel like an 18-year-old again.

What if I could go into suspended animation for eight years to stop aging until Carrie catches up with me and we're both physiologically 26? That would satisfy the terms of the legacy! No can do; long-term suspended animation is pure fiction, devoid of any scientific basis.

What if I could just slow down my own rate of aging for long enough so that Carrie and I would be the same physiological age at some point in the future?

Harry is a physics teacher, after all, and eventually the penny drops: Einstein! Special Relativity! Time Dilation! There's an answer there somewhere, he feels sure.

Harry pulls out his old physics texts, thumbs through them till he comes upon relativistic time dilation and the Lorentz equation that will enable him to calculate how fast and for how long he would have to travel in order to bring his physiological age close to Carrie's. He plays with the equation for a while, plugging in various numbers for speed and time to get a feel for what he would need to do.

He realizes that he would have to travel close to the speed of light for years in order to slow the flow of his personal time sufficiently. Not encouraging, but he doesn't give up.

Harry goes on-line to check on NASA's star survey program. Knows that they are running a series of manned near-light-speed probes to nearby stars. Checks the various mission specs for speed and duration of flights. Plugs the various time and distance numbers into the Lorentz equation and *whoa* – finds a near perfect match: a voyage to a white dwarf star named Chandra Centauri, the largest possible star of its type at 1.4 solar masses, just 4.9 light-years away, arriving back on Earth ten years hence, average speed 98% the speed of light. Ten years of Earth time but just two years of round-trip flight time for the ship's crew and passengers. Back on Earth when Carrie will be twenty-eight years old, both by her birth date and by physiology, and he'll be thirty-six years old by Earth time but only twenty-eight years old physiologically, the same as Carrie. Leaving in a month. Looking for physicists, biologists, and engineers to run the experiments. Perfect!

That wily old lady – she set us up! She knew I'd figure this out. So if I sign up for this, I'll be out of the picture for ten years while Carrie focuses on her studies without my distracting influence. Still, only two years of my time and we're back together again and we'll own that gorgeous house! Just by adding the word physiological, Grandma set things up so that Carrie would get a clean run at her studies and end up marrying someone her own physiological age **and** *get Grandma's gorgeous house. Nice triple. Glad I never played poker with Grandma!*

Calls Carrie and tells her what he proposes to do.

What. Have. You. Been. Drinking??

Nothing, Car. This is pretty straight-forward physics. NASA's been doing this for years now; it's perfectly safe.

Explain it!

*Sure, Car. My fellow-travelers and I take a shuttle to near Earth orbit where our spaceship awaits. The first part of our journey is spent accelerating to 98% of light speed, which astronauts refer to as .98***c***, "***c***" being the scientific symbol designating the speed of light. We spend the next year zipping to Chandra Centauri at that speed. On reaching its vicinity we slow down to a speed convenient for orbiting the star and taking various data on its magnetic field and the intensities and types of radiation it emits, which is the purpose of our mission. On completing our survey, we begin another acceleration phase to get us back up to .98c and head back to Earth at that speed. Approaching Earth we slow down to reduce our speed to Earth orbiting speed, about 20,000 km/hour, so we can dock with and transfer to our shuttle which takes us back down to Earth's surface – and our re-union!*

Since we'll be traveling at .98c for most of the journey, time will slow down for all of us in the spaceship by a factor of five, relative to you on Earth, so that when I arrive back I'll have aged only two years to your ten. At that stage, we'll both be physiologically twenty-eight years old, so we satisfy the terms of the legacy and we can marry!

But we'll be apart for the next ten years!

Yes, Car, ten years for you, but only two for me. That's the way it works. Nothing I can do about that, but I'll be making a big sacrifice too. I'll be stuck on a spaceship for two years with a bunch of strangers, and I'll be completely out of touch with what's happening on Earth for much of that time!

Wait a minute! I may not be a physicist, but I do know that nothing can travel faster than light. You will have travelled a distance of nearly ten light years in just two of your years. That's impossible!

No, Car. I will have travelled a distance of 9.8 light years in ten Earth years. But at the speed we'll be traveling, distances will shrink in our direction of motion, so as far as I'm concerned, the round trip distance to Chandra Centauri will be only 1.96 light years. The slowing down of time and the shrinking of distances in the direction of motion are both consequences of high speed motion predicted by special relativity.

Harry eventually persuades Carrie of the merits of his plan and he makes all the complex arrangements needed to be absent from Earth for the next ten years, including giving Carrie power of attorney over his business affairs so she can pay his mortgage, real estate taxes, etc. And he makes two more arrangements. Aside from being madly in love with Carrie, Harry is also a knowledgeable investor and a lover of fine wines, and he has every intention of taking full advantage of the journey he is about to undertake to Carrie's – and indeed, planet Earth's – future. So before he sets off, he buys some AAA bonds with a ten-year maturity and a case of good age-worthy California cabernet for their future wedding reception. He arranges for the wines to be stored in a cool cellar on Earth, knowing that they will be at their peak in about ten years – Earth years, that is. His bonds and wine investments were made in the full knowledge that in the mere two years of his high-speed space journey, both his Earth-based bonds and wine will age ten years – along with Carrie!

It may – or almost certainly *will* – be many decades or possibly even centuries before humankind learns how to build spaceships that are fast enough and well enough equipped to enable a future Harry to undertake a mission like this. The engineering challenges of near light-speed travel are daunting indeed, but they are just that – engineering challenges. Their eventual solution hinges mainly on the longevity of the human race or that of its evolutionary descendants. Nevertheless, the physics underpinning time dilation is now over a century old and has been validated extensively in physics laboratories and in clocks located in orbiting satellites.

So far, we have outlined Harry's imaginary trip into Carrie's future and claimed that it is logically possible under Einstein's special theory of relativity.

The rest of this book is aimed at justifying that claim.

LIMERICK TIME

On December 19th, 1923 the following limerick[6] appeared in Punch magazine:

There was a young lady named Bright
Whose speed was far faster than light,
She started one day
In a relative way,
And returned on the previous night.

That limerick, a tame example of the genre (though its opening line looked promising) was a humorous riff on the possibility of time travel arising out of Einstein's Special Theory of Relativity, which had been published eighteen years earlier. But Ms. Bright's hypothetical adventure is decidedly not possible under Special Relativity or any other scientific theory, since it involves traveling back in time and thus could violate a bedrock principle of causality, i.e., cause must precede effect. For if travel to the past were possible, she could, for example, travel back to a time before she was born and conspire to ensure that her father and mother would never meet, thereby annulling the possibility of her own future existence, so that she could never have travelled back in time in the first place to conspire to ensure that... Well, you get the idea.[7]

6. Named for a city in Ireland. Limerick was founded in 820 AD and has a population of about 150,000. It is home to a well-regarded university which has a strong focus on preparing students to compete and prosper in the real world that awaits them after graduation.
7. This causality problem is overcome in one interpretation of quantum mechanics by invoking the *many-worlds hypothesis*, by which the time traveler, having returned to a time before his parents had met and having successfully conspired to ensure that they would never meet, thereby instigates the creation of a parallel universe which proceeds forward in time without him. And if some readers consider that to be beyond crazy, well, welcome to the club. Nevertheless, it is a logically self-consistent hypothesis, and is the subject of considerable study within the physics community. In a limited sense it is possible to travel back in time by deploying what we learned in the Introduction about gravity-induced time dilation, and it involves none of the logical contradictions of Ms. Bright's journey. See appendix 8 (Traveling back in Time).

Apropos of our love story, here's a limerick that's truer to Einstein:

A physics professor named Harry
Fell in love with the much younger Carrie,
He went way out of pocket
For a ride in a rocket,
To fly to her future and marry.

The story being told in this limerick is entirely consistent with the possibility of time travel to the future predicted by Special Relativity. It is theoretically possible for its hero Harry to leave Earth and travel through space at speeds approaching the speed of light for an extended period of time, during which he would age fewer years than earthbound Carrie. Using the example from our love story, let Harry be twenty-six years old and Carrie eighteen years old on the day he leaves on his high-speed journey; by choosing his speed and travel time appropriately, he could arrange to arrive back on Earth on a day in the future *when he and Carrie are both twenty-eight years old.*

However counterintuitive it may seem, that claim is to be taken literally as it follows directly from the predictions and experimental proofs of Special Relativity. Let's assume that throughout their lives both Harry and Carrie had worn perfectly accurate watches that were set to zero the day each was born, and which kept track of days, months, and years as well as the usual seconds, minutes, and hours. In this story Harry's watch would read twenty-six years and Carrie's would read eighteen years when Harry begins his journey, *but both watches would read twenty-eight years on Harry's return to Earth.* In other words, while ten years will have passed for earthbound Carrie, only two years will have passed for Harry in his high-speed spaceship. And if Harry's spaceship had provided him appropriate diet, exercise, and human companionship, he would look and feel like a twenty-eight year-old *in all physiological respects.* Had he remained on Earth with Carrie, he would of course have been thirty-six years old and Carrie would be twenty-eight, and their watches would have so indicated.

The possibility of Harry's journey into Carrie's future will likely seem outrageous and an assault on common sense to many readers. Consider though, that our notions of common sense are just distillations of our individual life experiences, all of which are far removed from the circumstances in which time dilation effects begin to be noticed. Copernicus's discovery in the early 16th century that Earth revolves around the sun – and thus was not the center of the universe – was a huge assault on commonsense notions of that era. But mankind gradually assimilated that fact into its collective consciousness. So it will also, in the fullness of time, with Special Relativity.

Still, readers may reasonably object that it is impossible *by definition* for Harry ever to become the same physiological age as Carrie, since he was born eight years before her and therefore will always be eight years older. However, calculating people's physiological ages from their dates of birth is just a convention; it works fine since we earthbound slow-movers all age at much the same rate, not being exposed to the time-slowing effects of extremely high speed motion. Insurance companies assume this implicitly when writing life insurance policies. But one can imagine a future in which near light-speed travel has become so routine that on the day of our birth we will be required by law to be fitted with an internally embedded life-clock to keep track of our true physiological ages.[8]

In that eventuality, people's true (physiological) ages would be recorded by their embedded life-clocks and their actual dates of birth would be of diminished relevance, perhaps providing nothing more than an excuse for an annual birthday party.

The key point here, as will become clear later on, is that time dilation effects are about variations in the flow of time itself, whether recorded by clocks (mechanical, electronic, or atomic), or the rate at which a person ages, or by any other time-dependent process.

8. Such life-clocks could be powered indefinitely by tapping a tiny portion of the kinetic energy of blood flow or by utilizing the temperature difference between one's skin and an internal organ. This internal clock would transmit the measure of one's physiological age to an external device for display.

There's one more limerick that bears on our story:

There once was a fellow named Fisk,
Whose fencing was exceedingly brisk,
So fast was his action,
The FitzGerald contraction
Reduced his épée to a disc.

The provenance of this limerick[9] is somewhat uncertain, but it reflects (with just a teeny bit of exaggeration) another consequence of Special Relativity: length contraction. In 1889, in an attempt to explain the results of certain experiments with light, the Irish physicist George Francis FitzGerald proposed that objects contract in their direction

9. For some reason or other, relativity theory has inspired more than its fair share of limericks. Perhaps the difficulty of getting one's head around its counterintuitive ideas leads people to make light of the subject, and what more delightful way to do so than with a limerick or two? For readers who wish for more, the bibliography includes a reference to a superb book on relativity that's brimming with funny and clever limericks. But there's a price to pay: for every limerick in that book there's an obstacle course of at least a dozen mathematical equations on the way to the next limerick. But one can always cheat...

of motion by an amount dependent on their speed. The amount of contraction would be negligible at ordinary speeds but would increase rapidly as objects approached the speed of light. This was somewhat of an *ad hoc* proposal at the time, conceived solely to explain the results of certain experiments whose results were deeply puzzling, and FitzGerald offered no explanation for what the underlying contraction mechanism might be.

Nevertheless, the basic idea turned out to be correct; when Einstein introduced Special Relativity sixteen years later, length contraction exactly in accordance with FitzGerald's proposal was one of its predictions. As we shall see, time dilation and length contraction go hand in hand and both are unavoidable consequences of high-speed motion.

WHAT WE LEARNED IN THE INTRODUCTION

- That one of the consequences of Special Relativity is that the passage of time is not the same for every observer and in every location; specifically, that time slows down for people and objects in motion relative to people and objects deemed to be at rest, but that the effect only becomes significant at speeds approaching the speed of light.
- That this speed-dependent fact about the passage of time allows for travel to any desired future time, provided one has the technology to travel fast enough for extended periods.

Section I

IN WHICH WE INTRODUCE SOME ELEMENTARY PHYSICS

RELATIVE MOTION

Centuries before Einstein, the concept of relative motion was well understood, and since it plays a key role in Special Relativity we need to become familiar with it. It's a very straightforward idea and most readers will already understand it intuitively.

We now introduce Lauren and Liam, who will be involved in our various thought experiments. Liam is a curious but rather skeptical physics student who insists on verifying experimentally what he has learned in physics class.[10] He has persuaded his friend Lauren to be his accomplice in various experiments to help him understand relativity theory.

Imagine two cars being driven at a constant speed in the same direction on a smooth, straight two-lane highway on a moonless night, with headlights on. Car A is moving at 96 km/hour and car B, initially behind car A, is moving at 100 km/hour, according to their respective speedometers. The passenger of each car is seated behind the driver; Liam in car A and Lauren in car B. The road surface is very smooth and the car windows are closed, so Liam and Lauren are hardly aware that they are moving.

As Lauren's car draws level with Liam's, she looks out of her window and notices that her car is about to overtake his. She has a police radar gun to measure her passing speed, which as we'd expect, reads 4 km/hour. Further, knowing that her car is traveling at 100 km/hour (according to her car's speedometer) she deduces that Liam's car must be traveling at 96 km/hour.

10. As an aside, healthy skepticism is a valuable asset to scientists, as it can prevent them from being led astray by false notions of common sense or prevailing prejudices. But it can be taken too far. There's a story about a much-celebrated physicist who was having lunch with a friend. During the customary small talk, the friend remarked that it was a beautiful day. The physicist got up from his seat at the dining table, walked to the door and looked at the sky outside; then he returned to his seat, nodded to his friend and said: "Yes; it is."

We can reverse the situation by having Liam look out of his window and perform a similar calculation. His radar gun will measure Lauren's car overtaking his at a speed of 4 km/hour, and knowing that his car is traveling at 96 km/hour (according to his car's speedometer) he deduces that Lauren's car must be traveling at 100 km/hr.

What if both cars' speedometers were covered up so that Lauren and Liam had no idea of their own road speeds? Well, they would still be able to measure their passing speed as 4 km/hour, but that's all. For all they could tell, Lauren's car could be traveling at 50 km/hour and Liam's at 46 km/hour, or Lauren's at 72.3 km/hour and Liam's at 68.3 km/hour. Or – and here's the important point – Lauren's car could be traveling at 4 km/hour and Liam's could be stationary, or Lauren's could be stationary and Liam's could be traveling *in reverse* at 4 km/hour. Without speedometers to refer to, only the relative speed of their cars can be deduced.

If this story about Liam and Lauren had been told to Isaac Newton at the end of the 17th century, he might have asked how one could build horseless carriages that travelled at such speeds, but the concept of relative motion would have been well understood by him. His own experience of relative motion might have been gleaned as he woke up from a snooze and looked out the window of his horse-drawn carriage at the moment it was being overtaken by another, and momentarily thought that his own carriage was moving backwards; but a couple of centuries before Einstein he had already distilled such experiences into his famous three laws of motion.

The foregoing may be boiled down to this:

1. *All motion is relative;*
2. *there is no such thing as absolute motion; and*
3. *there is no location in the universe that can be said to be at rest and to which all other motion should be referenced.*

Earth rotates on its axis once per day; hence a point on the equator is moving at about 1670 kilometers per hour[11] relative to one of its poles.

11. Arrived at by dividing Earth's circumference at the equator (40,070 km) by 24 hours.

Earth orbits the sun at about 100,000 kilometers per hour. Our solar system orbits the center of our galaxy at even greater speed, and so on.

In our daily lives, when we say that a car or boat or train is moving at such and such a speed, it is taken for granted that the speed in question is relative to Earth's surface. But if we were traveling in a spaceship deep in interstellar space, there would be no obvious point of reference and thus no way of measuring our speed; indeed, the very concept of speed is meaningless without reference to some location or object.

Anyone who has traveled in an airplane in turbulence-free air will know that you can pour a glass of water as easily as if you were at home in your kitchen. The water will drop vertically *from your perspective* and end up in the glass, even though the airplane may be speeding forward at a thousand kilometers per hour relative to the air outside and the ground below; but so far as your sensory experience of pouring the water is concerned, the airplane could be stationary on the tarmac awaiting take-off clearance.

THE PRINCIPLE OF RELATIVITY

In the course of mankind's history of scientific research and discovery, in addition to a fondness for Occam's Razor[12] as a filter for explanations and theories, scientists have sought out laws of nature that have the widest applications and generality. They are aided in this effort by some empirical facts about our universe – *meta-laws,* we might call them – that simplify the task enormously.

The following three meta-laws may seem pretty basic, but our universe could conceivably have been constructed otherwise, perhaps by an inexperienced creator or one suffering a bad hangover while commanding "Let there be light!"

1. **The laws of physics do not change over time**
2. **The laws of physics are the same at every location in the universe**
3. **The laws of physics are independent of orientation relative to the universe**

There is abundant empirical evidence to support these meta-laws. For example, in the course of Earth's daily rotation about its axis and its annual orbit around the sun, the location and orientation of every point on Earth relative to the universe at large is constantly changing. Yet the laws of physics remain the same whenever and wherever they are tested. There is no such thing as Australian physics, or Italian physics, or Asian physics, or Caucasian physics. Scientists in all countries and of all religious faiths and ethnicities practice the same science and use much the same science texts, though expressed in different languages. And should we ever make contact with an alien civilization and learn how to communicate with them, there is every reason to think that their understanding of the laws of physics, including the mathematical structures needed to express them concisely, will be essentially identical to ours.

For an even more compelling example, we can observe the spectrum[13]

12. In modern parlance: "The KISS Principle": *Keep it Simple, Stupid!!*

of electromagnetic radiation (e.g., light) from a nearby galaxy. The Andromeda galaxy is around two million light-years distant from Earth, which means that it took the faint light from Andromeda two million years to travel the 20-million-trillion kilometer journey to reach our telescopes and spectrometers. Hence the light from Andromeda that we observe today originated within the Andromeda galaxy two million years ago. Yet the spectra of the various types of atoms that make up that galaxy are identical to the spectra of these same atoms on Earth and in our sun; and since the mechanism by which light is generated within atoms depends on several physical constants and several more physical laws, we are justified in concluding that the same physical constants and laws that apply here on Earth today also applied within the Andromeda galaxy some two million years ago and some 20 million trillion (20 quintillion) kilometers distant.

There is one more meta-law[14] that plays a crucial role in Special Relativity:

4. **The laws of physics are the same in different frames[15] that are in uniform motion relative to each other**

Note the qualifier "uniform." For our present purposes we are mainly concerned with uniform motion, i.e., where there is no acceleration or shaking that would lead you to conclude that you're moving. Uniform motion basically means motion in a straight line at constant speed.

---

13. A representation of the relative intensities of the various radiations from a star or galaxy by displaying them from left to right - from longer to shorter wavelengths - starting with microwaves, then infra-red, the visible colors, ultra-violet and on up to x-rays and gamma rays. A rainbow is a spectrum of *visible* radiation.

14. These four meta-laws are not formal laws of physics; you won't find them grouped and stated as such in physics texts, but they are nonetheless demonstrably true and are a useful way of introducing the fact of the constancy and universality of physical laws that underpins all of science.

15. Remember the concept of a "frame" defined earlier: in Special Relativity you will come across situations in which person A, who is deemed to be stationary, is observing person B, who is moving. A is said to be in the stationary or reference frame and B is said to be in the moving frame.

Here are a few examples of *non-uniform* motion: an airplane traveling in turbulent air; a car or train accelerating or braking or negotiating a bend; a spaceship accelerating in deep space. In all of these cases passengers would be able to tell they were moving even if the windows of their vehicles were blacked out.

When we speak of the laws of physics we mean all established scientific laws that govern the behavior of energy and matter, whether in the traditionally separate realms of physics, chemistry, or biology. The laws of chemistry and biology can, in principle at least, be derived from the laws of physics. Herein we use the term "the laws of physics" in its most general sense.

What do we mean when we say that the laws of physics don't change with time, location, orientation, or speed? Here is an example of a physical law: if a constant pushing force is applied to an object for a fixed period of time, the object will attain a definite speed that depends only upon its mass and the strength of the applied force (this is Newton's second law of motion). Here's another: suppose we have a flask containing a certain quantity of hydrogen and oxygen; we ignite the mixture and measure the total heat energy released using a device called a calorimeter; physicists (or chemists) would be able to accurately predict how much energy will be released in the process and they would do so using some of the laws of quantum mechanics and thermodynamics. One more: electric motors rely on the fact that a length of wire carrying an electric current will feel a force if placed between the poles of a magnet; the law quantifying this force is one of Maxwell's famous equations, of which we'll hear more later.

Those laws have been known for decades or centuries and they haven't changed over time; they work just as reliably in a laboratory on Earth as they would in a space ship traveling at high speed relative to Earth. That is the essence of the Principle of Relativity.

Some four centuries ago, Galileo Galilei noted that a heavy object dropped from the crow's nest of a ship traveling in smooth waters, irrespective of the ship's speed – provided only that it was constant while the object was falling – would always land on the deck *directly below the drop point,* just as it would if the ship were at rest. He may have been the first scientist to draw attention to this phenomenon. Centuries of scientific observations have since verified and expanded on Galileo's observation and today it is enshrined and made explicit in the Principle of Relativity.

By the early 1900's, centuries of contemplating relative motion and conducting experiments had led the scientific world to the grand conclusion that *all* laws of physics should be the same in all frames of reference in uniform motion relative to each other. Such frames are called *inertial frames,* implying that whatever movement they possess is by virtue of their own inertia, which is the same as saying that there are no external forces acting to alter their speed. Here are some examples of inertial frames: an aircraft traveling at constant speed in turbulence-free air; Lauren's and Liam's cars in our previous thought experiment; a train going at constant speed along a straight, smooth track; a stationary ship on a glass-like sea; any room in your house. More generally, you can be said to be in an inertial frame if you have no sensation of movement when your eyes are closed. Shaking, bouncing around, speeding around bends, acceleration or deceleration – any of these would disqualify it as an inertial frame.

The Principle of Relativity asserts that all experiments carried out in such frames will yield identical results. No exceptions to the Principle are known. In the year 1860, that Principle was – and still remains – scientific bedrock. Note however, that it has not been *proven* to be true. It is in the nature of a philosophical conjecture which squares nicely with scientists' ideals of how the world ought to be, and it is thoroughly justified by mountains of empirical evidence. Nevertheless, all it would take to refute the Principle is one confirmed counter-example. And, as we shall see, that almost came to pass.

PRINCIPLE OF RELATIVITY
SPECIAL THEORY OF "
GENERAL " " "

HOW MATERIAL OBJECTS, SOUND WAVES, AND LIGHT WAVES PROPAGATE

To gain a good appreciation of the problem Einstein solved, it will be helpful to understand the differences in how material objects, sound waves, and light waves are transmitted and received by stationary or moving observers. It is all quite straightforward, but really understanding it will make the problem that was driving physicists to distraction around the end of the 19th century much easier to follow.

A few thought experiments with Lauren and Liam will do the job.

Lauren is seated beside the driver of a train locomotive on a very long, straight section of railroad track. The window in front of her is open. She's brandishing a rifle with a muzzle velocity of 500 km/hour and she's an excellent shot. Liam is on a stationary flatbed car some distance in front of the locomotive and he has set up a standard marksman's target facing the locomotive. Behind the target's bull's-eye is a device for measuring the speed of incoming bullets. We will ignore the effects of air resistance and gravity on a bullet's speed (along with the fact that Liam would have to be nuts to stand anywhere near that target).

Lauren's locomotive is initially stationary. She fires a bullet at the target, scores a bull's-eye, and Liam measures its incoming speed to be 500 km/hour, as expected (Fig. 1).

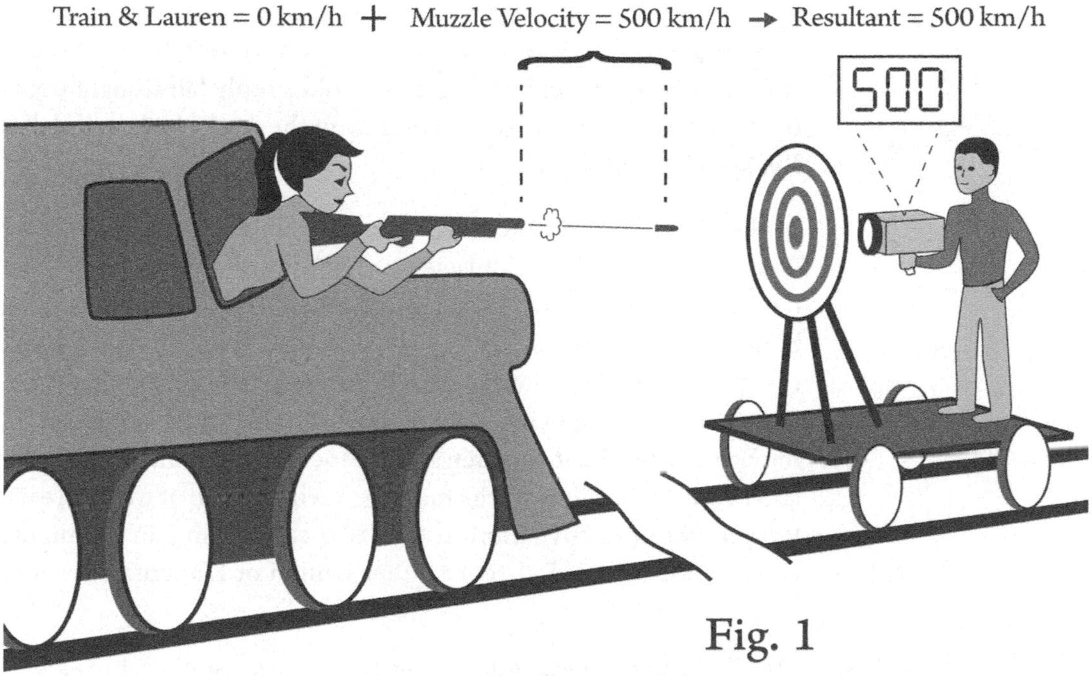

Fig. 1
Rifle and Target Stationary

The driver now backs up a few kilometers, and then drives the locomotive toward Liam's flatbed at 50 km/hour. Lauren fires another bullet and this time Liam measures its incoming speed at 550 km/hour (Fig. 2). This result should not be a surprise; since the locomotive along with Lauren and her rifle were all moving *toward* the target at 50 km/hour, and since bullets always leave the rifle's muzzle at 500 km/hour relative to the rifle itself, the locomotive's speed simply *adds* to the rifle's muzzle velocity and causes the bullet to travel forward at 550 km/hour *relative to the railroad tracks and Liam's stationary target.*

The driver next gets the locomotive going in reverse until it's moving backwards away from the target at 50 km/hour. Lauren fires a third bullet. This time Liam measures its incoming speed at 450 km/hour (Fig. 3). This result should not be a surprise either; since the locomotive along with Lauren and her rifle were all moving away from the target at 50 km/hour, and since bullets always leave the rifle's muzzle at 500 km/hour, the locomotive's speed simply *subtracts* from the rifle's muzzle velocity and causes the bullet to travel forward at 450 km/hour *relative to the railroad tracks and Liam's stationary target.* As an aside, if the locomotive were capable of going in reverse at 500 km/hour, the bullet on leaving the rifle would have zero forward speed *relative to the tracks and the target*; it would simply fall straight down onto the tracks as if someone standing on the tracks had dropped it! Really? Yes, really.

In this thought experiment with material objects (the bullets) we learned that if the locomotive is moving toward or away from Liam on his stationary flatbed he will measure the speed of the incoming bullet to have increased or decreased accordingly. We learned that the speed of the locomotive will add to the speed of the bullet when the locomotive is moving toward him and subtract from it when it's moving away.[16] What if Lauren and the locomotive remain stationary and Liam's flatbed does all the moving? Well, from what we've already learned about relative motion (the two cars passing in the night) and the Principle of Relativity, either Lauren or Liam may consider

16. This is true to very high accuracy at everyday speeds, but as we'll learn later, for objects traveling at sizeable fractions of the speed of light, simple addition or subtraction doesn't give the correct answer.

Train & Lauren = 50 km/h + Muzzle Velocity = 500 km/h → Resultant = 550 km/h

Fig. 2
Rifle Moving Toward Stationary Target

Train & Lauren = -50 km/h + Muzzle Velocity = 500 km/h → Resultant = 450 km/h

Fig. 3
Rifle Moving Away from Stationary Target

themselves to be at rest and that it's the other one who's moving. Thus, the answer depends only on whether the train and flatbed are moving closer to each other or are moving apart. It doesn't matter whether the locomotive or the flatbed or both are moving. If they're moving closer to each other the bullet's speed relative to Liam's target will increase and if they're moving apart it will decrease.

The foregoing is an example of how speeds combine in the case of material objects. If they act along the same line (i.e., if they are co-linear) as in the case of a bullet being fired directly out the front or back of a moving vehicle, speeds simply add or subtract as the case may be. (It may not be obvious to all readers that the speed of a bullet fired forward from a moving vehicle is simply the sum of the rifle's muzzle velocity and the vehicle's speed – *as measured by someone stationary outside of the vehicle*. There's an easy to follow proof of that in appendix 2, Composition of Velocities: Co-linear Case.)

We now consider the case of sound waves. This time Lauren has a device that emits a brief loud beep and Liam has a way of measuring the speed of that beep as it passes him on the flatbed. Here's what he will find. If both his flatbed and Lauren's locomotive are stationary, he will measure the speed of the beep as it passes his flatbed to be the normal speed of sound in air, which is about 1200 km/hour (Fig. 4). We're assuming that there's no wind blowing, so that the air (which carries the sound) is stationary with respect to the tracks.

Sound in Air = 1200 km/h + Liam's Velocity = 0 km/h → Resultant = 1200 km/h

Fig. 4
Sound Source and Detector Stationary

Fig. 5(a)
Sound Source Moving
Toward Stationary Detector

Fig. 5(b)
Sound Source Moving
Away From Stationary Detector

Suppose next that Lauren's device emits a beep when her locomotive is moving either towards (Fig. 5a) or away (Fig. 5b) from Liam's stationary flatbed, at say, 50 km/hour. In both cases Liam will still measure the beep to be passing him at 1200 km/hr. That's because his flatbed is stationary with respect to the air, through which the sound beep is traveling at 1200 km/hr..

Finally, if the locomotive is stationary and if Liam's flatbed is moving *toward* it at 50 km/hour he will measure the beep to be passing him at 1250 km/hour (Fig. 6a) and if his flatbed is moving away from the locomotive, he will measure the beep to be passing him at 1150 km/ hour (Fig. 6b overleaf). That's what we should expect since the beeps are traveling through the air at the constant speed of sound in air, but now Liam and his flatbed are moving through that air either toward or away from the oncoming beeps.

Sound in Air = 1200 km/h + Liam's Velocity = 50 km/h → Resultant = 1250 km/h

Fig. 6(a)
Detector Moving Toward Stationary Sound Source

Sound in Air = 1200 km/h + Liam's Velocity = -50 km/h → Resultant = 1150 km/h

Fig. 6(b)
Detector Moving Away
From Stationary Sound Source

The key difference between material objects (such as bullets) and waves (such as sound waves) is that material objects don't require a medium for their transport: they can just as easily travel through a vacuum as through air – more easily in fact, because air resistance just slows them down.

To appreciate how sound travels in air, think of a loudspeaker reproducing a pure "middle C" tone (which has a pitch of about 262 cycles per second). We can imagine the loudspeaker to consist of a simple disc-shaped plate of thin metal that is being moved back and forth (a millimeter or so each way) 262 times per second, by the audio amplifier that drives it. As the plate moves forward it slightly compresses the air directly in front of it and as it moves backward, it slightly decompresses the air directly in front of it. This alternating compression and decompression of the air directly in front of the moving plate, 262 times per second, is carried along through the air further from the plate; this constitutes a sound wave. Note that the speed of this wave through the air or other medium is largely determined by the medium's density. In a denser medium like water, sound travels about four times faster than it does in air. As one might expect, sound doesn't travel any distance in a vacuum since a vacuum, by definition, contains no material substance that can be compressed and decompressed.

As an aside, Liam will notice that the pitch of the beeps will always increase when the locomotive and flatbed are approaching each other and always decrease when they are receding from each other, regardless of whether the locomotive or flatbed or both are moving. This is the Doppler effect in action, the same principle that causes the pitch of a car horn or train whistle to change as the vehicle passes you. But changes in perceived pitch should not be confused with changes in the speed with which the sound beeps are traveling relative to the flatbed.[17]

17. There will be a change in the pitch of the sound beeps but not a change in their passing speed when the flatbed is stationary and the locomotive is moving toward or away from it; and there will be a change in both the pitch and the passing speed of the sound beeps when the flatbed is moving, whether or not the locomotive is moving. But if both locomotive and flatbed are moving at the same speed and in the same direction there will be a change in the passing speed of the beeps, but no change in pitch.

PROPAGATION OF LIGHT WAVES

As we've noted earlier, Einstein developed Special Relativity to account for the strange behavior of light waves, so let's now take a deeper look at the problem that his theory solved.

Back in Newton's time the very nature of light and the method by which it propagated were hotly debated and unanswered questions. One school of thought, championed by Newton, held that light was made up of tiny particles that behaved much like bullets fired from a gun and which would therefore travel unhindered through the vacuum of space. This was referred to as *the corpuscular theory.* Another school of thought held that rays of light were waves of some sort – *the wave theory*. Detailed experiments by Huygens, Young and others eventually proved conclusively that light had a wavelike nature, being subject to interference and diffraction much like sound waves or water waves. By the mid 1800s the scientific consensus favored a wave theory, which in turn begged the question: if light is a wave, how does it propagate and what is it that does the waving?

It was well known at the time that sound waves are pressure waves that propagate through either gaseous, liquid, or solid media, but not through empty space. Moreover, water waves need water to propagate, and the shear waves of an earthquake need solid ground. Thus, it seemed reasonable that light waves would also need some medium to transport them, so a medium called the luminiferous ether was conjectured to exist in order to allow for the propagation of light waves. This ether was *defined* as an invisible, weightless substance that permeated the entire universe of space and matter and that had a very specific set of physical properties to allow it to transport light waves. There was no way at the time to know if this all-pervasive ether (assuming it even existed) was stationary relative to Earth or the sun or to some planet of a distant star. But that didn't matter much; just as sound waves travel equally well in air that is moving relative to the source or listener, it was expected that light waves would travel equally well in the ether if it were moving relative to the source or observer.

Throughout the last few centuries, numerous attempts were made to measure how fast light travels and by the end of the nineteenth century its speed was known to be just a little under 300,000 kilometers per second in a vacuum and a tiny bit (90 km/sec) less in air.[18] This is about a million times faster than the speed of sound in air.

But some strange effects were observed...

We call on Lauren and Liam again, but now they are in deep space. Lauren is up front in a spaceship pointing a flashlight forward that emits a brief flash of light every second. Liam is in another spaceship in front of Lauren's, looking back toward Lauren and her flashlight. Liam has a device for measuring the speed of passing light flashes.

For the purpose of this thought experiment, we are going to assume that the ether, as hypothesized to explain the transmission of light, actually exists and that in the vacuum of space, light always travels through this ether at 300,000 km/sec. This was the prevailing view at the end of the 19th century. Also, for convenience, we are going to assume that initially both Liam and Lauren's spaceships are stationary relative to this ether and that they are a few kilometers apart.

When both spaceships are stationary, Liam will therefore measure the speed of each incoming light flash at 300,000 km/sec, the speed of light through the ether (Fig. 7).

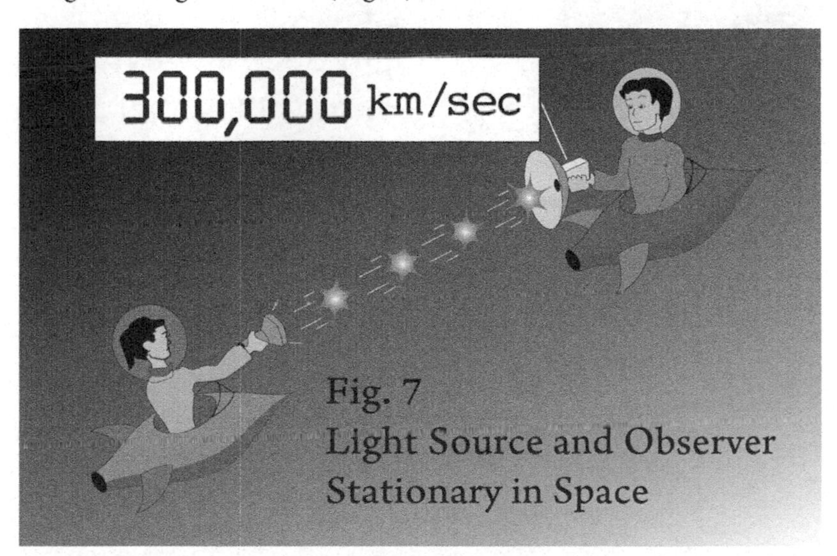

Fig. 7
Light Source and Observer Stationary in Space

18. The speed of light in water is about 225,000 km/sec, in ordinary glass it slows to 200,000 km/sec, and in diamond it slows even further to 124,000 km/sec.

Now, if Lauren fires up her engines and moves forward toward Liam (Fig. 8a) or moves backward away from him (Fig. 8b), Liam will still measure each flash to be passing him at the same 300,000 km/sec. That result would have been no surprise to Einstein's contemporaries for whom the ether was necessary to explain how light propagates. They would have reasoned that the light flashes were launched into the ether (in which by definition they travel at 300,000 km/sec) and since Liam's spaceship is stationary relative to the ether, he will measure the speed of the passing light flashes at 300,000 km/sec. Note the similarity to our previous experiment with sound.

Fig. 8(a)
Light Source Moving Towards Observer

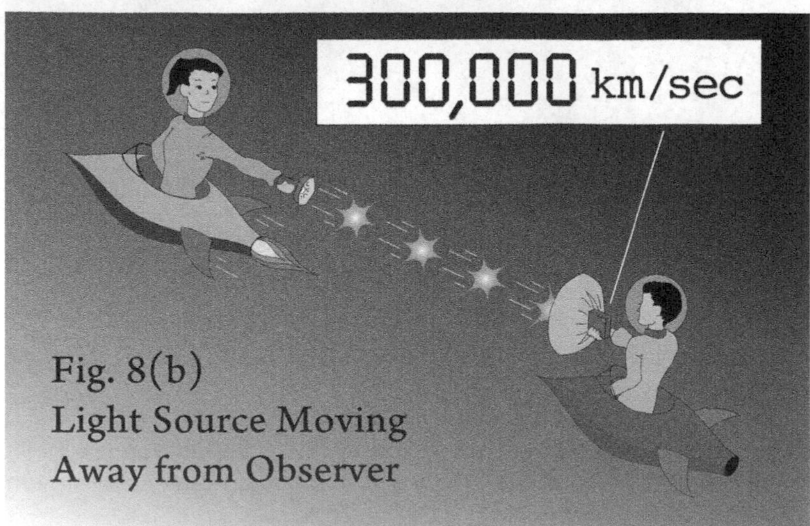

Fig. 8(b)
Light Source Moving Away from Observer

What should we expect if Lauren's spaceship remains stationary with respect to the ether and if Liam fires up his engines and begins moving either toward (Fig. 9a) or away (Fig. 9b, overleaf) from Lauren? A reasonable guess would be that Liam would measure the speed of the passing light flashes to be greater or lesser, respectively. That's because Liam's spaceship would now be moving through the ether, either toward or away from the oncoming light flashes which are traveling at constant speed relative to the ether. In other words, a reasonable guess would be that light – being a wave through the ether just as sound is a wave through the air – should behave as in the case of the sound beeps in our earlier train experiment with Lauren and Liam.

But that's not what happens. Irrespective of whether Liam's spaceship is moving toward or away from Lauren's and irrespective of the speed it's going at, Liam will always find the speed of the passing light flashes to be 300,000 km/sec.[19] Not only is this result different from the situation with the sound beeps, it's also different from the situation with material objects like bullets.

Fig. 9(a)
Observer Moving
Towards Light Source

19. The Doppler effect, which we discussed earlier in relation to sound waves has an analog with light waves. When the distance between a source of light and an observer is changing, whether caused by the motion of the source, the observer or both, the frequency (color) of the light waves changes, getting bluer on approach and redder on separation. This effect is of huge importance in astronomy and was a key factor in the discovery of the expanding universe.

Fig. 9(b)
Observer Moving Away from Light Source

It gets even crazier! Imagine that there are a whole bunch of spaceships spread out along a straight line in front of Lauren's spaceship and that they are all going at different speeds, some toward Lauren's spaceship and some away from it and that as Lauren's light flashes pass by each of these spaceships, their passing speeds are measured by the occupant of that spaceship. Incredibly, each occupant will measure the speed of the passing light flashes to be the same 300,000 km/sec. Some of the spaceships could be traveling toward or away from Lauren's at nearly the speed of light, yet the measured passing speed of each flash would still be 300,000 km/sec, not an iota more nor less.[20]

However strange and crazy it may seem (thought Einstein), light must behave just like that, otherwise it would be possible to catch up with and view a stationary light beam, and that seemed impossible given the nature of electromagnetic waves (the key aspects of which we shall discuss in Section II). In addition, the detailed experiments of Michelson and Morley performed at Case Western Reserve University in Cleveland, Ohio in 1887 strongly confirmed the constancy of the speed of light.

20. If we were of a mind to engage in a little magical thinking, we might speculate that the light flashes are intelligent and can figure out the speed of each spaceship they're about to pass, then adjust their speeds accordingly so as to always be measured as traveling at exactly 300,000 km/sec. But let's not go down *that* rabbit hole! Real physics is hard enough.

This was the puzzle that was engaging the physics community around the end of the 19th century and which finally was solved in 1905 with the publication of Einstein's Special Theory of Relativity, which we'll examine in Section II.

WHAT WE LEARNED IN SECTION I

- **That all motion is relative and there is no place in the universe that can be considered to be fixed. When we speak of speed, we must say what it is relative to.**

- **That the Principle of Relativity, which dates back to Galileo, states that the laws of physics are the same in all frames (vehicles, planets, etc.) in uniform motion.**

- **We learned how physical objects, sound waves, and light waves are propagated; that light behaves in unexpected and difficult-to-understand ways; in particular, that the measured speed of light is the same for all observers and is independent of the speed of the source of the light and independent of the speed of the observer.**

Section II

IN WHICH WE INTRODUCE EINSTEIN'S SPECIAL THEORY OF RELATIVITY AND CONSIDER ITS CONSEQUENCES FOR THE PASSAGE OF TIME

EINSTEIN'S SPECIAL THEORY OF RELATIVITY

Toward the end of the nineteenth century as we have seen, part of the scientific community was engaged in trying to understand the behavior of light. To their great surprise (and dismay) according to the experiments of Michelson and Morley (to be described later) and also certain astronomical observations, it appeared that the speed of light as measured by any and all observers was *independent* of how fast the observers were moving toward or away from the source of the light, in contrast with how sound waves or solid objects behave. We illustrated this strange property of light in our previous thought experiment involving Lauren and Liam in spaceships.

To get a better sense of how really strange that is, imagine that you're a traffic cop sitting in your car by the roadside in a 30 km/hour zone and that you're checking the speed of oncoming cars with your radar gun: suddenly you clock a car passing you at 60 km/hour. You fire up your engine and chase after it, savoring the prospect of writing a speeding ticket. But no matter how fast you chase after it, you find that you make no progress whatsoever, even when your own speedometer reads 100 km/hour. In frustration, you again point your radar gun toward the receding car and to your amazement you find that it is still moving away from you at 60 km/hour, when your expectation is that you should be rapidly catching up with it.

That's analogous to the situation scientists found when trying to measure the speed of light. Whether they were moving or stationary, light always moved relative to their instruments at the fixed speed of 300,000 kilometers per second; never more, never less. In the opening years of the 20th century, the scientific community was quite desperately trying to make sense of this.

A few decades earlier, as the American Civil War was raging, the Scottish mathematical physicist James Clerk Maxwell had devised a set of interlocking equations that unified the known laws of electricity and magnetism. A spectacular result of these equations was the *prediction* of a new kind of wave (later named *electromagnetic waves*) that should travel at a certain speed determined by just two well-understood physical constants associated with electricity and magnetism. The calculated speed was so close to the then accepted value of the speed of light that the inescapable conclusion was that light is a variety of electromagnetic wave. It has since been shown that gamma rays, x-rays, ultraviolet light, visible light, infra-red light, and microwaves – along with radio waves all the way from gigahertz frequencies down to very-low frequencies – are all electromagnetic waves differing only in frequency (or equivalently, in wavelength).

Maxwell's equations essentially laid the groundwork for our modern electronic communications technologies, and his discovery remains one of the greatest intellectual achievements of his or any other era, so much so that a Nobel laureate in physics, Richard Feynman, had this to say on the subject:

> *"From a long view of the history of mankind, seen from, say, ten thousand years from now, there can be little doubt that the most significant event of the 19th century will be judged as Maxwell's discovery of the laws of electrodynamics. The American Civil War will pale into provincial insignificance in comparison with this important scientific event of the same decade."*

That's as may be. But here was the rub. Maxwell's equation that specified the speed of electromagnetic waves was silent on a very important question: With respect to what do these waves travel at this colossal speed? From what we have learned already about all motion being relative, you can see that this was a very pertinent question.

If at the time you were still a proponent of the corpuscular theory of light as championed two centuries earlier by Newton – and many scientists were, despite what appeared to be strong evidence against it – a reasonable guess would have been as follows: light beams are emitted from their sources (e.g., flashlights) at a fixed speed – the speed of light – just as bullets are emitted from *their* sources (rifle barrels) at the rifle's muzzle velocity. In other words, a reasonable guess at that time would have been that the "muzzle velocity" of a flashlight or other source of light is simply the speed of light. If that were the case one would expect light to travel at different speeds relative to observers depending on whether the source of the light was moving toward or away from them, just as the bullets did in one of our earlier thought experiments.

But that guess was suspect because light was known by then to have wave-like properties and that's not how sound waves behave, as we've seen. Sound waves travel through air and other media at a specific speed regardless of the speed of the device emitting them. So the notion of light acting like bullets from a gun was untenable. It was expected therefore that light waves launched into the ether from a flashlight or other source would travel at 300,000 km/sec, relative to this ether, regardless of the speed of the source of the light relative to the ether.[21]

As a consequence, two people traveling at different speeds relative to the ether would get different answers if they measured the speed of light, since the speed of light would be relative to the ether. Therefore, since the speed of light is a law of physics derived directly from Maxwell's equations, it would no longer be true that the laws of physics appear the same in all frames of reference in uniform motion relative to each other. If true, that spelled doom for the universality of the Principle of Relativity.

Nonetheless, since the ether hypothesis did answer the question – *the speed relative to what?* – and since it also provided the medium for the propagation of electromagnetic waves (which seemed essential at the time), by the end of the nineteenth century the ether looked like it was going to be a permanent fixture in physics, the sacred Principle of Relativity be damned.

In the mid-1880s, two American physicists, Albert Michelson and Edward Morley, designed an elegant and precise series of experiments to measure Earth's speed relative to the (presumed to exist) ether. Their experiment, in essence, measured the speed of light in the direction

21. The development of quantum mechanics in the early part of the 20th century, by Planck, Einstein, Bohr, Heisenberg, and others, led to the present view that light can behave either like a particle or like a wave, depending on the experiment being carried out. Indeed, Einstein was awarded the Nobel prize in 1921, not for his Special or General Theories of Relativity, which were then still controversial, but for his elucidation of the photoelectric effect, in which light behaves like a stream of particles, called photons.

of Earth's travel in its orbit around the sun, in which it is traveling at the substantial speed of about 100,000 km/hour. They conducted this experiment at different times of the year when Earth was in different locations in its orbit and therefore traveling at different speeds and directions relative to the ether. The experiment was repeated many times and was known to be sensitive enough to detect very small differences in the speed of light, but the answer was always the same: the measured speed of light was unaffected by Earth's speed through the ether. The experiment, which was fully expected to confirm the presence of the ether, caused a crisis in the world of physics.

Many different ideas were advanced in attempts to explain this null result. One of the least crazy was the notion that Earth dragged the ether along with it in its orbit, thus making it effectively stationary with respect to Earth's surface where Michelson and Morley were conducting their experiments. This would explain their null results, but there were other problems with this idea, which need not concern us here.

Enter Albert Einstein, all of twenty-six years old at the time. He had a deep and largely self-taught knowledge of contemporary physics, and as a teenager he had closely studied Maxwell's equations. He knew that these electromagnetic waves were caused by a changing magnetic field inducing a changing electric field in its immediate vicinity, this in turn inducing a changing magnetic field in *its* vicinity, and so on, resulting in a wave traveling forward at the speed of light. He imagined himself in deep space accelerating closer and closer to the speed of light trying to catch up to such a wave. As he caught up to it (in his mind's eye) he deduced that it would manifest strangely as vibrating electric and magnetic fields that were stationary in empty space. He felt sure that such a thing would be in conflict with Maxwell's very successful theory. Yet he could not imagine what might stop him from catching up to a light ray if he traveled fast enough, and even if he could catch up to it, what he expected to observe would make no sense. This puzzle nagged at him on and off during his late teens and early twenties.[22]

22. Lest readers think that Einstein must have been the original nerd to be worrying about such matters at that young age, he was in fact quite the rake. His anti-establishment views and Bohemian

In early 1905 Einstein was one of many scientists deeply engaged in trying to understand the apparent fact that all measurements of the speed of light gave the same answer irrespective of the speeds of the source or of the instrument doing the measuring. His teenage thought experiment about chasing after a light ray was never far from his mind. On one occasion, he was so frustrated by his lack of progress he was about to give up on the problem altogether. He decided to visit his best friend and scientific sounding-board, the engineer Michele Angelo Besso, and in the course of telling Besso about the problem, he suddenly saw the key to its solution.

Einstein's epiphany was "simply" this: He noted that physicists of the time, himself included, were tying themselves in knots trying to understand how light could behave so strangely, using prevailing notions of common sense. In particular, they were wedded to the assumption that the explanation rested in the realm of Newtonian physics, where time and space were considered as independent and absolute entities. He decided instead to accept the strange behavior of light as a fact about the universe and see where it led. He began with the following assumptions:

1. **That all physical laws are the same for observers in uniform motion (the Principle of Relativity).**
2. **That the speed of light through empty space is independent of the speed of its source, as measured in any frame of reference.**

These assumptions, taken together, imply that one could never view a stationary light beam, or even one going at any speed other than the nominal speed of light. Einstein appreciated that this constancy of the speed of light would be strange indeed compared to how sound

---

lifestyle would have made him quite inconspicuous had he lived in San Francisco seventy years later during the Summer of Love! But he also had a deep passion to understand the laws of nature and that passion co-existed in the same brain with a visceral disrespect for authority and a deep suspicion of handed-down dogma and "common sense" - traits that were to serve him well in his chosen profession. Eventually, however, the gods took their revenge for his apostasy. In his own words: *"To punish me for my contempt for authority, fate made me an authority myself."*

waves or material objects like bullets behave, but that's what mother nature seemed to be shouting from the rooftops.

In 1905 he sent a paper to one of the most respected scientific journals of his time – *Annalen der Physik*. The title of his paper, written in his native German, translated to *On the Electrodynamics of Moving Bodies*. It began with those two assumptions, which today are called *the Postulates of Special Relativity.* It continued with various thought experiments involving two frames in relative motion. After solving some fairly simple equations, he ended up with a series of strange predictions about time, length, and mass, which later became known as *Einstein's Special Theory of Relativity.*[23]

His theory *predicted* that:

1. **Time passes slower in moving frames relative to stationary frames.**
2. **Moving objects appear to contract along their direction of motion, and equivalently, distances appear to shrink in the direction of motion of the traveler.**
3. **Events that appear simultaneous to stationary observers do not appear simultaneous to moving observers; and both viewpoints are equally valid.**
4. **The momentum of an object increases with its speed and approaches infinity as its speed approaches the speed of light.**
5. **Mass is a form of energy according to the equation $E = mc^2$.**[24]
6. **The speed of light is the limiting speed in the universe. No material object, wave, or information can travel faster.**

It wouldn't be entirely true to say that the theory was met with a big yawn, since many of the great physicists of the day saw it for what it was – a solution to the dilemma with which physics was grappling,

23. The "Special" designation arose years later after Einstein had extended the theory to include situations involving acceleration, in what came to be known as his General Theory of Relativity. Since his 1905 theory dealt only with situations involving uniform motion, it became retrospectively known as the 'Special Theory', special in the sense of "restricted to non-accelerating frames".

24. Where E is the equivalent energy of a body of mass m, and c is the speed of light.

and a major revolution in how we view the universe. It did not garner widespread acceptance for two reasons: its teachings were highly counterintuitive and difficult to believe, and at the time there were no ways to test their validity since they dealt with situations involving either unattainably high speeds or that required clocks considerably more accurate than clocks of that era.

In the intervening years, all the predictions of the theory have been verified to great accuracy, largely due to dramatic increases in the precision and resolution of (atomic) clocks. Some, like the equivalence of mass and energy (which provides the basis for nuclear reactors and nuclear weapons) have dominated international politics. Others, like time dilation and length contraction, have changed our understanding of the universe at the deepest levels.

Of particular relevance to the subject of this book is item 1, which falsified the common-sense notion that the flow of time is the same for everyone. In so doing it provided the scientific basis for travel to the future, as we shall see.

The Special Theory of Relativity is now bedrock physics; it underpins many branches of science and technology and is essential to a deep understanding of subatomic processes, since subatomic particles routinely travel at speeds close to the speed of light.

Nevertheless, Newton's three-century-old laws still play a dominant role in science and engineering and there is every reason to think they will do so for millennia to come. With the exception of problems involving extreme speeds – in excess of tens of thousands of kilometers per second – or the minuscule dimensions inside atoms, or the intense gravity in the region of black holes, Newton's physics can describe and predict events with great accuracy. So when the popular press opines that Newton's laws of physics were completely overthrown by Einstein's laws, it is well to remember that in the hundred years or so since Einstein introduced his Special and General Theories of Relativity, jet aircraft dominated the skies, men were sent to the moon and returned safely to Earth, and that it was largely Newton's laws that provided the scientific bedrock on which every one of those achievements was based.

Let's now take a closer look at what is meant by Einstein's discovery that time passes slower in moving frames relative to stationary frames.

NEWTON'S TIME AND EINSTEIN'S TIME

Prior to Einstein, as we have learned, the reigning theories in physics were largely based on the work of the seventeenth century English scientist Isaac Newton. Implicit in the Newtonian worldview was the conviction that space and time are completely separate entities; that space is the three-dimensional fixed and immutable stage on which the events of our lives and of the universe unfold; that time is an entirely separate dimension by which we experience and measure the sequential unfolding of these events; and that its rate of passage is the same for everybody and in every location in every city, planet, or galaxy throughout the universe.

Indeed, that is still the worldview of just about everybody today whether they have thought about it or not. Why? Because it is a very accurate description of what we perceive with our unaided senses.

Consequently, it was thought to be obvious that if perfectly accurate clocks were located throughout the universe and were initially synchronized to each other, they would continue to tick away at the same rate and would ever after display the same time, regardless of whether they were stationary or in motion. (We are not concerned here with the type of clock, or its construction, or its reaction to being dropped or shaken, or exposed to extremes of heat or cold. We assume perfectly accurate idealized clocks that can faithfully record the passage of time *in their immediate vicinity*.)

As we have learned, one of the results of Special Relativity was to show that this everyday conception of time is false and that it is an illusion resulting from our greatly restricted experience of our universe. Specifically, Einstein showed that the rate of passage of time is relative rather than absolute, and in particular that *time passes more slowly in moving objects relative to objects deemed to be at rest.*

Before proceeding, we need to be absolutely clear about what's being asserted in the previous paragraph: Consider that the universe is full of objects (and people) moving at different speeds and in different directions relative to each other. Here on Earth, people may be sitting still or walking or running. Cars, trains, planes, and ships are moving hither and thither at various speeds relative to Earth and to each other. Earth is progressing in its annual orbit around the sun, and the entire solar system is in orbit around the center of our local galaxy, the Milky Way, which in turn is orbiting the local cluster of galaxies, which in turn is moving within a super-cluster of galaxies, and so on. Now imagine that there are perfectly accurate clocks distributed throughout this vast universe of moving objects. Choose any one of those clocks as your reference clock, and imagine that you are standing beside it. *From your point of view, the clocks in every location in motion relative to this clock will be running slower; and since the business of a clock is to record the passage of time in its vicinity, this means that time itself must be passing more slowly in vehicles or regions in motion relative to you, our deemed-to-be-stationary observer.* Moreover, it doesn't matter whether the motion of a particular clock you're observing is toward or away from your own clock or is moving at some other angle to it: its relative speed is the sole factor that determines the amount of slowing down of time you will observe. We demonstrate this in appendix 4 (Independence of Time Dilation on Direction of Motion).

Furthermore, the situation is fully symmetrical. If person A situated at location A notes that person B's clock at location B is running slower by a certain amount by virtue of B's motion, then person B will note that person A's clock is running slower relative to her clock, and by the same amount. This may seem counterintuitive, but if you ponder it for a moment you will see that it follows directly from the Principle of Relativity, which allows either A or B to consider themselves to be stationary and the other to be moving. Nonetheless it is reasonable to ask: How can A's clock be running slower than B's and at the same time B's be running slower than A's? We shall address that question later in this book.

It bears repeating that we are using clocks in these thought experiments just because they are the traditional devices used for recording the passage of time. But we could equally well rely on any other time-based phenomenon, such as the rate of a person's heartbeat if it were sufficiently regular. For example, if we are observing a clock that ticks once per second held in the hand of a person whose heart beats once per second, then any observed slowing of the clock caused by the time-dilation phenomenon we're discussing here would be accompanied by an identical slowing of the person's heartbeat. Time dilation is all about the slowing down of time itself; clocks are just the traditional devices we use to keep a record of its passage.[25]

These previous paragraphs may seem very puzzling or even downright untrue based on your personal experiences of traveling in cars or aircraft. You *know* that you don't have to re-synchronize your watch with your house clock every time you've been out for a drive or returned from a flight to somewhere or other. (We ignore time zone changes in this discussion as those are just artificial rules to ensure that 12:00 pm corresponds more or less to mid-day everywhere on Earth.)

The reason for the discrepancy between your own experiences and these facts about time dilation is simply this: at the speeds we encounter in our daily lives, whether walking or traveling in cars or aircraft, the amount of time dilation incurred is far too small to notice; a precision clock placed on an aircraft traveling at 900 km/hour from San Francisco to London and back would, at the end of the 20-hour journey, read only about 24 *nanoseconds* less than an identical clock that remained on the ground. That's a difference of just one part in three trillion; the only clocks that could measure time that accurately are the atomic clocks found in a small number of specialized physics laboratories.[26]

So why, readers may ask, make a big deal about such an absolutely trivial difference in the passage of time? Because the effect, while miniscule even at the speed of a jet aircraft, grows rapidly larger as objects approach the speed of light: at about 86.6% of the speed of light time slows down by a factor of two; at 99.5% of the speed of light time slows down by a factor of ten; at 99.995% it slows down by a factor of 100. And at the actual speed of light, time comes to a

25. Special Relativity uses the term *Proper Time* to refer to the passage of time on stationary observers' clocks, and the term *Co-ordinate Time* to refer to the inferred passage of time in a moving frame.
26. Those clocks rely for their accuracy on the fact that the frequency of the light emanating from certain atoms (usually Cesium) is predictable and stable to an extraordinary degree.

complete standstill! Appendix 5 contains a graph showing how time slows with increasing speed; it shows that as you approach the speed of light, a small further increase in speed dramatically increases the amount of slowing down.

So, for time dilation to become noticeable, objects must move at an appreciable fraction of the speed of light. Today, our fastest spacecraft reach speeds of about 30 km/sec, which is only 1/10,000th the speed of light. At that speed, the clocks on a spaceship traveling to Mars and back on a six-month round trip journey would lose less than one-tenth of a second due to time dilation.

We'll ask you to accept these strange facts about time for now. A major purpose of the remainder of this book is to demonstrate why they are true, and show how they open up the possibility of travel to the future.

RELATIVISTIC TIME DILATION VS. THE DOPPLER EFFECT

As we have learned, relativistic time dilation is a subtle but real effect that goes unnoticed in our daily lives but becomes quite obvious and of great importance for objects traveling at speeds approaching the speed of light. The Doppler effect, in contrast, is a physical principle that can create an *illusion* of time dilation (and its opposite, time contraction), but has a different cause. Before proceeding, it will be helpful to explore the difference.

To do so, we introduce another thought experiment: Liam is standing on a railway platform near the rear of a parked train. Lauren is in the train's rear looking out the back and operating a device which emits a loud beep exactly once per second as timed by her watch – and also by Liam's watch *while the train is stationary*. The train now pulls away from the station and quickly accelerates to a speed of 34 meters per second (m/sec), which is about 1/10th the speed of sound in air. As the train recedes from Liam, he will find that the time he measures between beeps has increased to 1.1 seconds. Has time slowed down for Lauren and her fellow passengers? Is this an example of relativistic time dilation? Not at all. What's happening here is that each beep travels to Liam's ears at the speed of sound in air, which is about 340 m/sec, but for every second that passes, the train is 34 meters further away from him, so the next beep takes an extra 1/10 of a second to reach his ears. Hence the arrival time between beeps has increased to 1.1 seconds and it may seem to Liam like a slowing down of the rate at which Lauren's device emits beeps. But it isn't. This is just another manifestation of the Doppler effect, which causes the pitch of the horn of a fast-approaching train or car to suddenly decrease as it passes you; this effect was first explained in 1842 by the Austrian scientist Christian Doppler.

If Lauren's train were to stop and instantly reverse back toward Liam at the same speed, he would measure the time between beeps at only 0.9 seconds,[27] because for every second that passes the train would be 34

27. There will of course be some delay before Liam notices the beeps to be arriving faster, depending on how far away the train is when it goes into reverse.

meters closer to him, so the sound of the next beep would take 1/10 second less to reach him. Hence the arrival time between beeps would decrease to 0.9 seconds and it may seem to Liam like a speeding up of the rate at which Lauren's device emits these beeps. Furthermore, if he had counted the number of beeps since the train started moving, by the time the train reached him at the end of its return journey the total number of beeps would be the same as the number of seconds taken for the round trip, according to his watch. Hold that thought!

Now imagine a similar experiment carried out in deep space. Liam is standing at a space station near the rear of a parked spaceship. Lauren is at a window at the back of the spaceship operating a light which emits short bright flashes at the rate of one flash per second as timed by her watch, and also by Liam's watch *while the spaceship is stationary*. Lauren's spaceship now pulls away from the space station and quickly accelerates to a speed of 30,000 kilometers per second, which is 1/10th the speed of light. As the spaceship recedes from Liam, he will find that the time he measures between flashes has increased to 1.1 second. Has time slowed down for Lauren and her fellow passengers? Is *this* an example of relativistic time dilation? Again, no. What's happening here is that the light flashes are traveling toward Liam at the speed of light, which is 300,000 km/sec. But for every second that passes, Lauren's spaceship is 30,000 kilometers further away from him, so each flash takes an extra 1/10th of a second to reach him. This again is another example of the Doppler effect, but applied to light rather than sound.

If Lauren were to stop her spaceship and reverse back toward Liam at the same speed, he would then notice that the time between flashes had *reduced* to 0.9 seconds,[28] because for every second that passes the spaceship would be 30,000 kilometers closer to him and each flash would take one tenth of a second less to reach him. And if he counted the number of flashes since the spaceship started moving, by the time it reached him at the end of its return journey the total number of flashes would be the same as the number of seconds taken for the round trip, according to his watch. Hold that thought also!

28. As in the earlier case of sound, there will also be some delay before Liam notices the flashes to be arriving faster, depending on how far away Lauren's spaceship is when it goes into reverse.

Now, about those thoughts we asked you to hold: In both experiments we ended up comparing the total number of beeps or flashes recorded by Liam with the total number of seconds taken for the round trip, according to his watch. Why? Because that is precisely equivalent to comparing the total elapsed time as measured by Lauren (on the train or on the spaceship) with that measured by Liam who remained at the stations. And if both these times were the same at the end of these journeys, as indeed they were, then there has been no time gained or lost by Lauren during her journeys. Simple as that! (Lauren and Liam could also confirm this by comparing the times indicated on their watches, which would of course read the same.)

The point of introducing the Doppler effect into our discussion on relativity is that an observer listening to or watching events unfold in a vehicle moving away from (or toward) him, as Liam did, will hear or see those events unfold at a slower (or faster) rate than if the vehicle were at rest relative to him. And this might seem very similar to the time dilation effects between moving objects predicted by Einstein. But the underlying mechanisms are entirely different. The Doppler effect manifests only when the distance between observers is changing, whereas relativistic time dilation occurs whenever observers are moving in any direction relative to each other, even if their separation remains the same. For example, if Lauren's spaceship travelled in a circle with Liam at the center, no matter how fast she moves there would be no Doppler effect because the distance between her and Liam would not be changing. But there would still be relativistic time dilation. Notice also, as explained earlier, that relativistic time dilation (by definition) always involves an observed slowing of time, never a speeding up. In the case of Lauren's and Liam's experiment, when their vehicles were moving closer to each other each would observe the other's time processes to *appear* to be speeding up. Note also that at the end of Lauren's journeys the time indicated by her watch will always be the same as that indicated by Liam's, so there is no net time gained or lost; it's all just an illusion, caused by the finite speeds of propagation of sound and light.[29]

29. In the interests of accuracy, we should point out that because the spaceship is traveling at 10% of the speed of light in this experiment, a small amount of relativistic time dilation would kick in, though it

To underscore that point, imagine that Liam could both hear and see events going on in Lauren's train as it receded from him and that Lauren has both devices, one that makes sound beeps and the other that makes light flashes, both at the rate of one beep and one flash per second, occurring in sync. To do so, he might deploy a very sensitive directional microphone and a telescope pointed at the receding train. He'll still find that the beeps from the train are arriving 10% less frequently than normal (since the train is receding at 1/10th the speed of sound) but the light flashes from the train now appear virtually instantaneous because the train's speed is an absolutely miniscule fraction of the speed of light. So what he observes is that as the train recedes further and further away from him the flashes and the beeps become more and more out of sync, due to the fact that the speed of light is about one million times greater than that of sound in air.

We can now state clearly what is meant by relativistic time dilation: Let two people first synchronize their watches and let one of them undertake a journey involving appreciable speeds and time; if on getting back together again their watches are no longer synchronized, then time dilation has occurred. In all cases the watch of the traveler will indicate less elapsed time than the watch of the stay-at-home person. If the speeds are our common earthbound speeds, the amount of time dilation will be so small that their watches will be unable to parse the difference. But if the speeds are an appreciable fraction of the speed of light and the journey lasts for months or years, the resulting time dilation can be hours, weeks, years and even centuries! In the latter case such time travelers could arrive back and meet with their own great, great, great, great grandchildren.

---

would amount to only about one twentieth of the apparent slowing down of time caused by the Doppler effect. Even in the previous train example, there would also be a relativistic time dilation effect, though it would be even more minuscule and well below our ability to measure because of the small speeds involved. There is always a time dilation effect whenever one object is moving relative to another, whatever their relative speed. However, it is not caused by sound or light transmission time delays as objects move away from you or toward you. These, as we have seen, manifest as Doppler effects.

Some readers may have been jolted awake by the previous paragraph. Up until now we've been at pains to point out that the Principle of Relativity allows each person to declare that she's the one who's stationary and that it's the other person that's moving. That being so, you might ask, what is the distinction between the traveler and the stay-at-home person that causes the traveler to age more slowly, since each can claim (under the Principle of Relativity) to be stationary while it's the other person that's moving? Here's the short answer: The Principle of Relativity, as we've explained earlier, applies only to people and objects in uniform motion, where no acceleration is involved. But for the traveler to get anywhere she would need to accelerate to get up to speed and decelerate again to stop. This asymmetry, when fully explored and analyzed leads to the definite conclusion that the traveler will have aged less than the stay-at-home person. Nevertheless, this Twin Paradox,[30] as it has come to be known, has caused much confusion to scientists and non-scientists alike. We will resolve this "paradox" later.

In the event that some readers missed this critically important point, we'll state it again: Thought experiments about time dilation often invoke someone in a train or spaceship moving relative to an outside observer, who is deemed to be at rest. The observer is usually said to be in the stationary or reference frame and the person in the vehicle is said to be in the moving frame. The object of the thought experiment is to figure out what the person in the stationary frame can deduce about what's happening in the moving frame, and/or vice versa. Both are provided with perfectly accurate clocks. If, for example, Liam concludes that Lauren's clock is running slow relative

30. There seems to be two quite separate meanings of the Twin Paradox, depending on who's writing about it. One focuses on the mere fact that one of the twins will have aged less than the other and because this is contrary to our established beliefs about time, it's said to be a paradox. The other meaning focuses on the fact that it's always the traveling twin who ages less and this *seems* (at first glance) to be in contradiction to the Principle of Relativity by which either twin may consider herself to be stationary and the other twin to be moving. The first meaning is trivial; we adopt the second meaning in this book as the resolution of that version of the paradox is instructive.

to his, this observation is not just about Lauren's clock; it's about all time-dependent events in Lauren's frame: her heart rate; the rate at which her hair and nails are growing; the rate at which she is aging; the rate at which the cup of coffee on her desk is cooling, etc.

Furthermore, though Liam may conclude that all time processes in Lauren's frame are running slow relative to his, from Lauren's point of view everything is normal. Her heart rate etc., *as timed by her clock*, is unchanged. The rate of growth of her hair and nails, as timed by her clock, are unchanged. Her innate sense of the passage of time is unchanged. This follows directly from the Principle of Relativity.

ANOTHER THOUGHT EXPERIMENT

Our earlier train experiments with Lauren and Liam were designed to explain the difference between how material objects, sound waves and light waves propagate. We did so by showing how the measured speeds of bullets, sound waves, and light waves are affected by the speeds of the objects emitting them and by the speeds of the devices measuring those speeds. For simplicity and convenience, we arranged the motion of the sources and observers to be along the same trajectories that the bullets, sound waves and light waves were traveling.

In Section III we will be considering a thought experiment that makes a direct connection between the speed of a vehicle and the slowing down of time for people or clocks within that vehicle, as measured by a stationary observer. That experiment will show how Einstein's two Special Relativity postulates inevitably imply time dilation. Everything we have learned up to now is aimed at helping us fully appreciate the details and conclusions of this upcoming experiment.

But before we get to that, we first need to learn how an object behaves when it has speed components that are at right angles to each other.

Here goes: Liam is on a railroad platform facing a railroad track. Lauren is in a carriage on a train approaching Liam's location from his left at a speed of 20 m/sec, Fig. 10(a). Below the train's ceiling is a low-powered rifle with a muzzle velocity of 15 m/sec. It is pointed directly downward and its muzzle is exactly 3 meters above the floor of the carriage. There's also a hole in the floor directly below to allow the bullet to exit the carriage. When Lauren's carriage is directly opposite Liam, she presses a button that fires the rifle. We will ignore the effects of gravity and air resistance.

From Lauren's perspective within the carriage, the bullet travels straight downward toward the floor of the carriage and it exits through the hole in the floor. It takes 1/5 of a second to do so (3 meters divided by 15 m/sec.) This is obviously what will happen when the train is stationary and, both from our own experiences of traveling in trains

What Lauren Sees Fig. 10(a)

(or other vehicles) and also from the Principle of Relativity, we know that the same thing will happen in a train traveling at constant speed along a straight track.[31 (overleaf)]

From Liam's location on the platform, he sees things differently. He also sees the bullet exit the rifle muzzle and disappear through the hole in the floor 1/5 of a second later, just as Lauren did. But in that 1/5 of a second, the train moved to his right a distance of 4 meters (20 m/s times 1/5 of a second). So the actual path of the bullet *from his perspective* is as shown in Fig. 10(b) and the length of that path is exactly 5 meters.[32 (overleaf)] Fig. 10(b) shows four snapshots of the bullet on its way from the rifle's muzzle to the hole in the floor of the carriage.

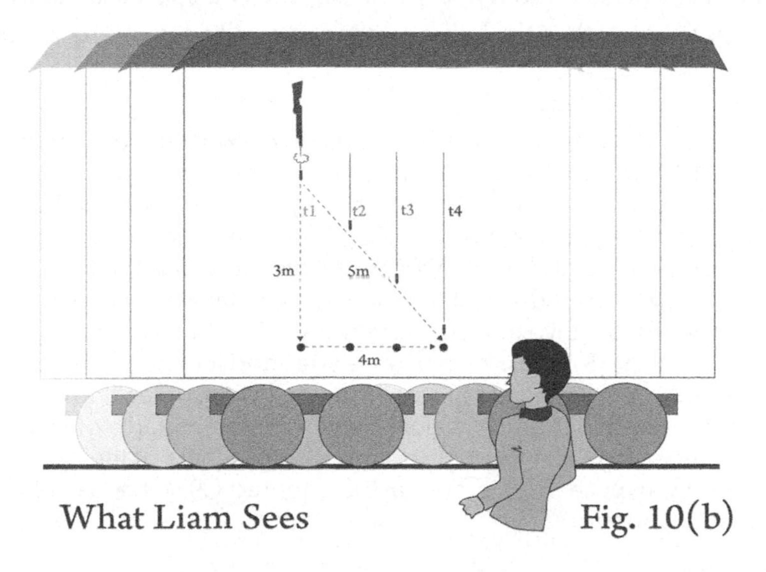

What Liam Sees Fig. 10(b)

Now all this is centuries-old physics and if you ponder it a little, it will make sense. But if you thought for example that the bullet should still have dropped straight downward *from Liam's perspective,* wouldn't it hit the floor 4 meters to the left of the hole since the train had moved that distance during the bullet's flight? It would therefore have punched a second hole in the floor where it hit and that would really be crazy, because Lauren saw no such thing from within the carriage. The explanation is that from Liam's perspective, when the bullet left the rifle's muzzle it had two speed components; the rifle's muzzle velocity that ejected the bullet vertically downwards and the train's horizontal speed that the bullet possessed before it left the rifle's muzzle; and both speed components were maintained (there being nothing to stop them) until the bullet buried itself in the ground beneath the train. Thus, the diagonal path of the bullet as seen by Liam.

Note that the speed of the bullet *from Liam's perspective* has increased from 15 m/sec to 25 m/sec, which is why it could cover the 5-meter diagonal distance in the time it took to get from the ceiling to the floor of the carriage. This is another example of combining speeds, but there's a difference. Recall that in the earlier train experiment with Lauren and Liam, the speed of the train was either added to or subtracted from the rifle's muzzle velocity to get the speed of the bullet relative to the tracks. That's because the two speeds were acting along the same line (co-linear), i.e., both were parallel to the railroad track. But in the present thought experiment, the two speeds are at right angles to each other, the bullet moving vertically downward and the

---

31. How do we know this from our own experiences? Imagine you're sitting in a train facing backwards and enjoying a nice cup of tea while the train is moving. If you drop your cup you know that the tea will land in your lap, not in the lap of a passenger sitting directly opposite and facing you! We should give eternal thanks for the natural laws that underpin the Principle of Relativity; without them, travel would devolve into constant brawls with fellow passengers.

32. It's exactly 5 meters because we arranged it to be so for numerical simplicity. Readers may remember the Pythagorean theorem for a right-angled triangle, in which the sum of the squares of the two sides enclosing the right angle is equal to the square of the opposite side - the hypotenuse. In our case 3 squared (9) + 4 squared (16) equals 5 squared (25).

train moving horizontally. In this case, instead of simply adding or subtracting them to get the resultant speed and direction of the bullet, we need to use a different method, which is really quite simple. Draw two lines at right angles to each other; make the lengths of two lines proportional to the two speeds to be combined; complete the rectangle; the length and direction of the diagonal is the resultant speed and direction of the bullet. For readers who'd like a fuller understanding of this, see appendix 3 (Composition of Velocities: Orthogonal Case).

The important takeaway from this experiment – and also from our earlier experiment in which Lauren fired a bullet out the front of a moving train – is that the speed of a material object (in both those cases, a bullet) relative to an observer can be affected if the device that ejected the bullet (the rifle) is itself moving, whether along the path of the bullet or indeed at any angle relative to the path of the bullet. In this experiment, the speed of the bullet relative to Liam had increased from the 15 m/sec muzzle velocity of the rifle to 25 m/sec, as a result of the train moving at 20 m/sec at right angles to the path of the bullet relative to the rifle.

In contrast to this behavior of material objects, we have learned that the speed of light is always the same relative to any and all observers and will not be affected by the speed of the light's source. As a result of this empirical fact, our next thought experiment forces us to conclude that time passes more slowly in moving frames, relative to frames deemed to be stationary.

WHAT WE LEARNED IN SECTION II

- That Einstein developed his Special Theory of Relativity in order to come to grips with the strange behavior of light.
- That his Theory predicted that the passage of time and the measurement of length are both speed dependent; that the speed of light is the limiting speed in our universe; and that matter is a form of energy.
- The manner in which the passage of time depends on speed.
- What the Doppler effect is and how it can create an illusion of time dilation.
- The difference between relativistic time dilation and the Doppler effect.
- How to combine speeds that are operating at right angles to each other.

INTERMISSION

IN WHICH WE TAKE STOCK OF WHAT WE'VE LEARNED SO FAR, AND OF THE ROAD AHEAD

REVIEW TIME

We have now covered enough history and elementary physics to allow us to follow Einstein's thinking and get to the bottom of time dilation. We have learned:

- **That all speed must be defined relative to something else and that the concept of absolute speed (or absolute motion) has no meaning**
- **That the Principle of Relativity asserts that the laws of physics remain the same whether tested in a uniformly moving frame (vehicle), or in a frame deemed to be stationary (like in a ground-based laboratory)**
- **How to combine the speeds of moving objects (e.g., trains and bullets) that are traveling along the same line (co-linear) or that are at a right angle to each other (orthogonal)**
- **How material objects (e.g., bullets), sound waves, and light waves travel from place to place**
- **That light always travels at the same speed irrespective of the speed of its source and irrespective of the speed of any and all observers who are measuring its speed**

We have also learned that this last point describes the conundrum that puzzled physicists around the end of the nineteenth century. That, unlike physical objects and sound, the measured speed of light is completely unaffected by the speed of sources emitting the light or the speed of observers measuring it.

And we have learned that Einstein realized that this constancy of the speed of light could not be accounted for within the Newtonian worldview, where time and space were believed to be unchanging entities independent of any and all external influences. So, he took the constancy of the speed of light as a law of nature, and he devised some thought experiments to see what its consequences might be. The result was his Special Theory of Relativity.

So, here's the plan for Section III:

We will examine another thought experiment that's quite similar to the most recent one (in which Liam stood on a platform observing a passing train within which Lauren fired a bullet from ceiling to floor). But we'll move the experiment to outer space to give us some room and we'll replace the train with a spaceship and the bullet by a stream of light pulses (flashes).

And most importantly, we'll insist that our observer (Liam) reports his findings taking full account of the Principle of Relativity and the empirical fact of the constancy of the speed of light.

We will discover that he must logically conclude that time in Lauren's spaceship is passing slower than the time indicated by the clocks on his platform.

We will state the mathematical formula that relates the slowdown of Lauren's time (relative to Liam's time) to the speed of her spaceship. We show how to derive that formula in appendix 5 (Proof of the Lorentz Formula).

So, by the conclusion of Section III, we will have shown how Einstein deduced that the passage of time is relative rather than absolute, and why time always moves more slowly in vehicles or frames that are in motion relative to an observer deemed to be at rest. And we'll know how to calculate the amount of time dilation that occurs at a given speed, and readers can use that to confirm the time dilation numbers that applied to Harry in our love story.

Section IV demonstrates that the journey Harry took would indeed cause him to age at one-fifth the rate of stay-at-home Carrie, so that when they re-united they were the same physiological age. In so doing it proves that the infamous Twin Paradox is not a paradox at all, but rather a misapplication of the Principle of Relativity.

SECTION III

IN WHICH WE SHOW HOW THE PRINCIPLE OF RELATIVITY, COUPLED WITH THE CONSTANCY OF THE SPEED OF LIGHT, GIVES RISE TO TIME DILATION

TIME DILATION THOUGHT EXPERIMENT

The topics we've covered up to now were chosen to provide readers with a good understanding of the problem Einstein solved, along with enough basic physics to fully appreciate his solution. In this section, we apply all we have learned to see how and why time dilation happens for moving objects.

We have learned that in the case of physical objects and sound waves, their measured speed would be affected by the motion of the source or the observer in a manner fully in accordance with the laws of physics as understood toward the end of the 19th century, and indeed, with common sense also. But light behaved differently; both the results of the Michelson-Morley experiments and a careful parsing of Maxwell's equations pointed to the speed of light being the same for all observers, irrespective of their own speeds or the speed of the device emitting the light.

It would be inaccurate to say that Einstein solved the problem of the constancy of the speed of light, as it was not a problem in and of itself. The real problem was that the physics of his day could not account for it. And his Special Theory of Relativity showed how it could be fully and elegantly accounted for, though he had to abandon the prevailing assumption that the experience of time and space are the same for all observers, whether they're moving or stationary.

Einstein approached the problem as a mathematical thought experiment in which he imagined two frames, one deemed to be stationary and the second moving at a constant speed relative to the first. He considered what an observer in the stationary frame would deduce about objects in the moving frame, taking into account the Principle of Relativity and the constancy of the speed of light, the two assumptions (postulates) on which he based his theory.

The most elegant and simple thought experiment ever devised to illustrate time dilation asks you to imagine two plane mirrors, one lying flat with its reflective side facing directly upwards, with the second mirror some distance (30 centimeters, say) directly above it, with its reflective side facing directly downwards. The mirrors may be mounted on each end of a glass tube or held in place by a frame of some sort. Now imagine a single light photon bouncing up and down continuously between these two mirrors. Since light takes one nanosecond to travel 30 centimeters, this device can be considered to be a clock which "ticks" every two nanoseconds, each tick being the time it takes for the photon to travel up to the top mirror and then back down to the bottom mirror. This device (though impossible to construct in practice[33]) is known as a "light clock" and is deployed in many books on relativity to show why time dilation occurs for moving objects, as follows:

Readers are first invited to consider such a clock when it's stationary relative to an observer and therefore ticking at the same rate (once every two nanoseconds) according to the observer's clock. The observer "sees" the photon traveling vertically upwards and downwards at the speed of light. Then the clock is imagined to move horizontally past the stationary observer at, say, half the speed of light. From the observer's perspective, the path of the photon is now like a saw-tooth in shape and therefore longer than the simple vertical up/down path when the clock was stationary. But since the speed of light is the same for all observers, the observer must infer that the longer path the photon must now traverse between each "tick" causes the moving clock to tick at a slower rate than when it was stationary, thereby illustrating the fact of time dilation for moving objects. It's a matter of simple geometry to calculate the length of the photon's saw-tooth path for a given horizontal speed and thereby arrive at the Lorentz equation for time dilation.

This is a brilliantly simple and clever thought experiment. But whereas trained scientists who frequently have to analyze far more abstract situations would have no difficulty accepting this thought experiment as clear and convincing proof of the reality of time dilation (given

33. Unless the two mirrors are precisely parallel to each other, the photon will very quickly veer off in one direction or another and miss one of the mirrors completely, thereby stopping the clock. Moreover, mirrors are not perfectly reflective, so even if a light clock deployed a burst of light comprising millions of photons, they would all be absorbed by the material of the mirrors in a very short time.

the Principle of Relativity and the constancy of the speed of light) it requires a fair amount of faith on the part of readers not used to such abstractions to accept its conclusions.

So here we present an equivalent thought experiment that accomplishes the same end. Though a bit more involved, it should be more convincing to readers, particularly those of a Missourian[34] disposition for whom accepting the conclusions of the light clock thought experiment might be a bit of a stretch. *What? You expect us to believe that time slows down for people in motion based on the antics of a single photon bouncing between two mirrors in a thought experiment that can't actually be performed?*

Though the previous light clock experiment and the upcoming thought experiment both show that observers in stationary frames must logically *infer* that time in moving frames runs slower than time in their own frames, by means of a follow-on thought experiment we will show that such inferences have the same factual value as direct observations or measurements.

As we proceed, we will come across a situation in which Liam makes a measurement or does a calculation that produces a counter-intuitive result and the reader might be tempted to think, "Well, it's all fine and dandy that the various rulers and clocks gave that particular answer, but what's really going on here?"

Bear in mind that our entire knowledge of what we call our universe was created by our brains interpreting inputs from our five senses (six if you insist), often aided by various instruments (clocks, tape measures, cameras, telescopes). If every observation or measurement we make about some particular aspect of the universe yields the same answer, then we are justified, absent evidence to the contrary, in calling that answer "a fact about our universe." In our case, if a careful set of observations and/or measurements leads us to conclude that time is moving slower in some other location relative to our own location, then so be it. However, since a conclusion like that would defy common sense for many readers, one may be tempted to ask, "Is that

34. License plates on Missouri-registered vehicles refer to the state as the "Show Me" state, reflecting the storied skepticism of its people..

really what's happening here?" Ask such a question of a physicist and you are likely to get one of two responses: "Here's an accurate clock and ruler; go measure for yourself." Or: "Can you offer an alternative explanation that fits all the facts?"

Keep that in mind as we proceed, because chances are you will be tempted! So here we go...

Lauren is a passenger in a spaceship which contains a laser that can emit extremely brief, green-colored light pulses (flashes) (Fig. 11a). The laser is mounted on the ceiling of the spaceship, and it is pointed directly downward at a small glass "porthole" in the floor, through which the light flashes exit. The laser's output end is exactly 3 meters above that porthole. Note the similarity to our last train experiment.

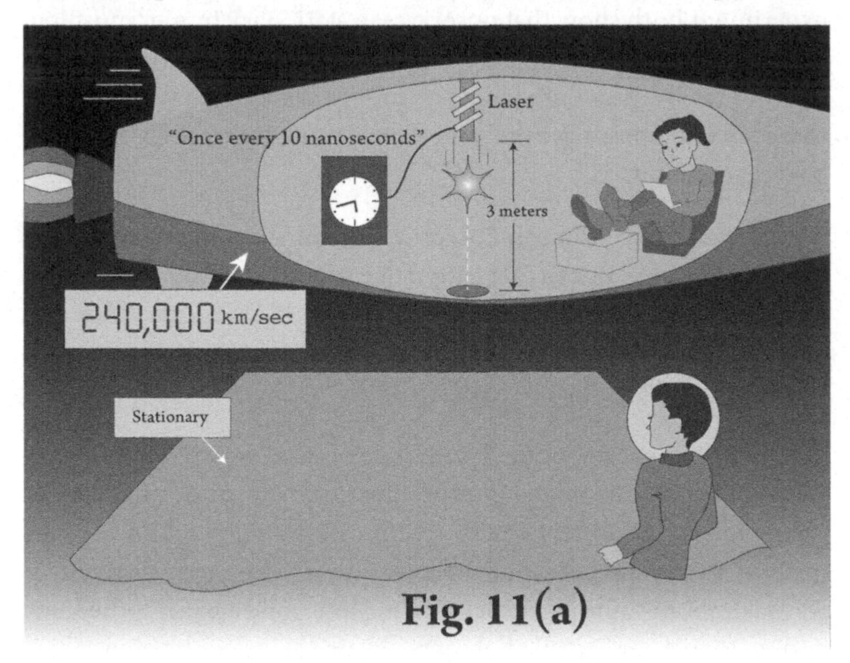

Fig. 11(a)

The laser is triggered from the spaceship's master clock, which is extremely stable and accurate, and which "ticks" every 10 nanoseconds, each tick causing the laser to emit a new flash. As we know, light travels 3 meters in 10 nanoseconds; that means that at the precise moment that one of the flashes exits the porthole, another one exits the laser.

The same precision clock that triggers the laser also drives a wall clock in Lauren's spaceship to tell time. In addition, Lauren is wearing an accurate wrist watch which runs independently of the spaceship's master clock. Therefore, any change in the rate of that master clock (e.g., caused by malfunction) would equally affect the wall clock, but not Lauren's watch.

Liam will be on a space platform making measurements and observing the experiment. Initially Liam joins Lauren in the spaceship to check out the equipment. He measures the distance from the laser to the porthole and confirms that it is exactly 3 meters. He compares the ticking rate of the ship's master clock with his own precision clock and notes that they are fully in sync.

Liam returns to the platform and the spaceship's captain backs the spaceship up a (really) long way off to the left of the platform. He then drives the spaceship at full power toward the platform, getting up to 80% of the speed of light (relative to the platform). On reaching that speed, he shuts down the engines in order to maintain that speed as the spaceship passes Liam on the platform.[35]

Being a careful observer, Liam had previously set up a device to confirm the spaceship's speed relative to the platform. That device simply clocks the time it takes for the front edge of the spaceship to travel a certain distance along the platform. His precision clock measures that time and he had previously used a precision ruler to calibrate the distance. As the spaceship comes in line with the platform, his instruments confirm its speed to be .8c (80% of light speed).

Liam's goal in this experiment is to figure out what must be happening to the flow of time in the speeding spaceship from his perspective as a stationary observer, i.e., he wants to be convinced of the reality of time dilation. There are no rockets attached to his space platform and even if it happened to be drifting relative to some distant planet or star, he would be completely unaware of it (The Principle of Relativity). Therefore, he may (and does) consider his space platform to be stationary in the vast darkness of space.

35. It would take about a year for the spaceship to attain that speed if the captain kept the acceleration at about 1 g (the same as on Earth), but not having to be concerned with inconvenient details like that is the beauty of thought experiments.

He knows of the empirical fact that the speed of light as measured by all observers is a fixed 300,000,000 m/sec and that it is not in the least affected by the speed of its source (Einstein's second postulate). He also knows that from his perspective, as the spaceship passes by him the light flashes he observes through the spaceship's window must follow a diagonal path from the laser to the porthole since the spaceship is moving rapidly to his right while the flashes are moving vertically downward from top to bottom of the spaceship's cabin. So, his first inescapable conclusion is that the path of the light flashes *from his perspective* must be longer than 3 meters, being the hypotenuse of a right-angled triangle with a 3-meter vertical side. Elementary geometry: see Fig 11(b).

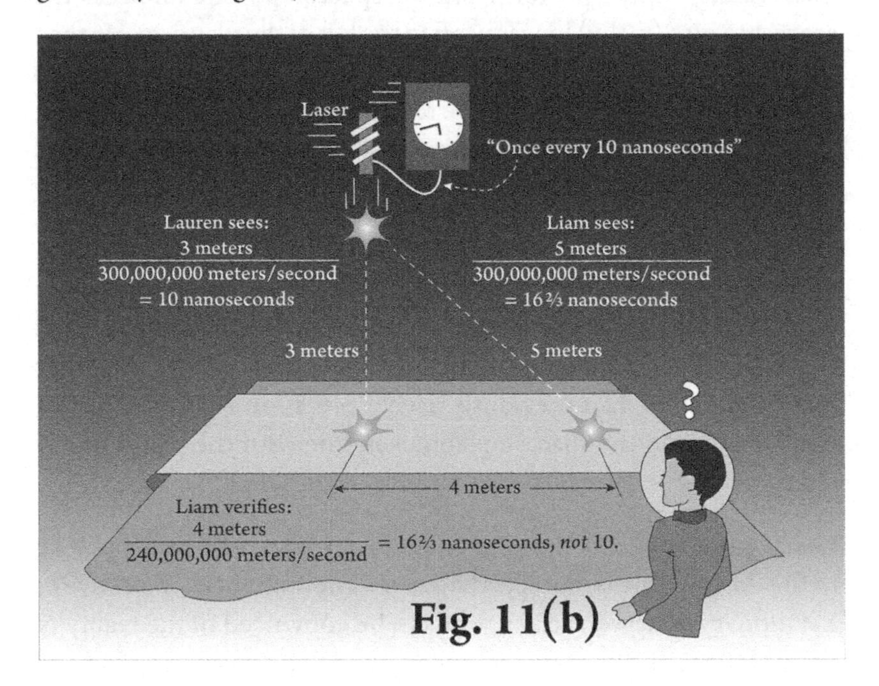

Fig. 11(b)

His second inescapable conclusion is that since it takes the flashes 10 nanoseconds to travel the 3 meters to the porthole from Lauren's perspective inside the spaceship, it must take them longer than 10 nanoseconds *by his platform clock* since they now have to travel more than 3 meters, but at the same speed.[36]

In order to calculate the amount of time the flashes must take to reach the porthole by his clock, Liam only needs to figure out the distance they have to travel, since he already knows that their speed is the unvarying speed of light. He also knows that the spaceship will have travelled 80% of that same distance horizontally since it's traveling at 80% the speed of light, and he knows that the flashes will have travelled vertically by exactly 3 meters, the distance between the laser and the porthole. So, to get the answer, he needs to know the lengths of the horizontal side and the hypotenuse of a right angled triangle whose vertical side is 3 meters and whose horizontal side is 80% the length of its hypotenuse. Using Pythagoras' theorem again, he finds the answer: the horizontal side is 4 meters and the hypotenuse is 5 meters. That means that the flashes travel 5 meters from the laser to reach the porthole (from his perspective) and that the spaceship travels 4 meters to his right between light flashes.

Let's consider this result for a moment. Liam had previously confirmed that the laser in the spaceship is mounted exactly 3 meters above the porthole and that it emitted a light flash every 10 nanoseconds as measured by the spaceship's master clock when the spaceship was stationary. His own platform instruments measured the passing speed of the spaceship at 80% of the speed of light (240,000,000 meters per second). Relying on nothing more than the empirical fact of the constancy of the speed of light for all observers, he calculates that each flash has to traverse 5 meters (from his perspective) to get from the laser to the porthole. That means that each flash must take 16 ⅔ nanoseconds by his clock (5 meters divided by 300,000,000 meters per second) to reach the porthole. That's 6 ⅔ nanoseconds *more* than the time it took when Liam measured it earlier while the spaceship was parked at the platform.

Liam's conclusion? That time in the speeding spaceship *must* be running 40% slower than his platform time, meaning that for every minute his clock advances, the spaceship's clock advances by just 36 seconds. He wonders if Lauren is feeling light-headed since her heart-rate is

---

36. Compare this to our previous thought experiment with Lauren in the train carriage; though the path of the bullet was also longer from Liam's perspective on the train platform, its speed was increased by the train's horizontal speed, so it took the same time by Liam's clock to exit through the hole in the floor as it would if the train were stationary.

down to 60% of its normal rate (by his clock).

Not a bit! Lauren is inside the speeding spaceship, feet up, watching Dangerous Liaisons on her laptop, and all is right by her. Just because Liam is out there on the platform furiously calculating and coming to some strange conclusion about how slowly time is passing in her spaceship is of no concern to her (the Principle of Relativity, again). Her watch is still ticking away at the same rate as the clock on the wall, which we know is controlled by the same master clock that triggers the laser. She would still measure the pulses exiting the porthole every 10 nanoseconds by her clock if she could be bothered to get up from her comfortable armchair to do so. Her heart is actually beating *a little faster* than usual (a hot scene in the movie) according to both the clock on the wall and her wristwatch, and she feels just fine; thank you very much for asking.

We have now reached a major milestone in our quest to understand the physics of time travel: We have carried out a thought experiment to examine the consequences of applying Einstein's two relativity postulates to a situation involving relative motion. The result is the inescapable conclusion that time passes more slowly for objects in motion *relative to observers deemed to be at rest*. The objects in this case are Lauren and her spaceship and the deemed-to-be-stationary observer is Liam. This is exactly the same conclusion Einstein arrived at, though using a more theoretical approach as befitted the broad and grand goals he had set out to accomplish.

Even though Liam was careful to a fault in designing this experiment he's still a little skeptical of his conclusions. *"Is this really what's happening?"* he wonders. *"I couldn't actually measure the time between the light flashes exiting the porthole since Lauren's ship passed by me in about a millionth of the blink of an eye."*

A reasonable doubt for sure, given the radical nature of his conclusion.

Though a doubt we anticipated:

About that little glass porthole that lets the light flashes out – the one some readers might have been wondering about? You get that in the last train experiment Lauren wouldn't have wanted the bullets to be shattering the floor of her train carriage and showering her with wood splinters; hence the hole in the floor to let them out to safely expend their energies in the ground below. But light pulses from an ordinary laser? The most they'd be likely to do, absent the porthole, is to warm the spaceship's floor a bit. So, why the porthole? Glad you asked. *Because it allows us to confirm Liam's inference that time must be moving at a slower rate in the speeding spaceship relative to his platform time.*

Here's how: In place of where one might expect to find tracks if this were a railway platform instead of a space platform, we laid a very long sheet of light-sensitive paper pulled taut and glued in place; and we arranged for the captain to fly the spaceship just a little distance above that paper as it traversed the platform. So, now that the spaceship has gone its way, we can take a look at that paper. You see, each laser flash exiting the spaceship's porthole will make a mark on the light-sensitive paper directly below the porthole, so let's measure the distance between marks. Look! it's exactly 4 meters!

Knowing the distance between marks and the speed of the spaceship, we can now calculate the time interval between those flashes. Divide the distance between the marks by the speed of the spaceship (4 meters divided by 240,000,000 meters per second) and we get 16 ⅔ nanoseconds for the interval between flashes. This is a direct measurement and thus provides empirical proof that the spaceship's clock, which triggered those flashes every 10 nanoseconds when it was stationary relative to Liam, is now doing so every 16 ⅔ nanoseconds *according to Liam's platform clocks*. It follows directly that time within the speeding spaceship must be running 40% slower than space platform time.

Now, we're going to do a little mathematics, or more accurately, we're going to preview the result of a little mathematics which we'll actually do in appendix 5 (Proof of the Lorentz Formula).

If Liam were to generalize his calculation of how long the light flashes must take to get from the laser to the porthole *from his perspective*, he would come up with the following formula. $T_1 = T_2 / \sqrt{1 - k^2}$ where T_1 is Liam's assessment of the time it takes the light pulses to get from the laser to the porthole according to his clock, T_2 is Lauren's assessment according to her clock, and k is the fraction of the speed of light at which the spaceship is traveling. This is the famous Lorentz formula for relativistic time dilation. In our thought experiment, if you put in 10 nanoseconds for T_2 and 0.8 for k, you will find that T_1 comes to 16 ⅔ nanoseconds.

Before moving on there are a few points about this thought experiment we should note.

We got a front row view of the Principle of Relativity in all its awesomeness. There was Liam observing the spaceship zooming by at 80% of the speed of light relative to his space platform, while for Lauren inside, relaxing in a comfortable chair and enjoying a movie, everything seemed completely normal, so normal in fact that her spaceship might as well have been at rest for all she could tell without looking out a window. Indeed, by this same Principle of Relativity, if she had set up some experiment to figure out what was happening on *Liam's* space platform, she would have concluded that time on Liam's platform was running 40% slower than her spaceship's time. Why? Because, Liam and his space platform appear to her – and by the Principle of Relativity *in effect are* – speeding by her *stationary* spaceship at 80% of the speed of light.

All very well, you might say, for both Lauren and Liam to consider themselves to be at rest while the other is moving. We get that, since you've been drumming the Principle of Relativity into our heads throughout this book; but how can it be that Liam infers that Lauren's clock is running 40% slower than his own clock, while at the same time Lauren infers that Liam's clock is running 40% slower than her own clock? That would seem to be a logical impossibility!

Appealing again to the Principle of Relativity, we could answer this by asking: "How could it be otherwise?" If, for example, we had said that Lauren's clock appears to be running slower than Liam's while Liam's clock appears to be running **faster** than Lauren's, at first blush readers might be inclined to think that there's no contradiction there and move on. But wouldn't that beg this follow-on question? "Given the fact that all uniform motion is relative, and that therefore both Lauren and Liam can each consider themselves to be at rest and the other to be moving, how could they possibly come to *different* conclusions?" Answer: they logically could not. So, we're faced with the seemingly paradoxical fact that each must conclude that the other's clock is running slower than their own clock. And that requires some explanation, which we'll provide by examining what appears to be an even more paradoxical situation and showing that there's no paradox whatsoever involved.

Consider this thought experiment: Observer O is standing on a space platform in interstellar space (Fig.12).

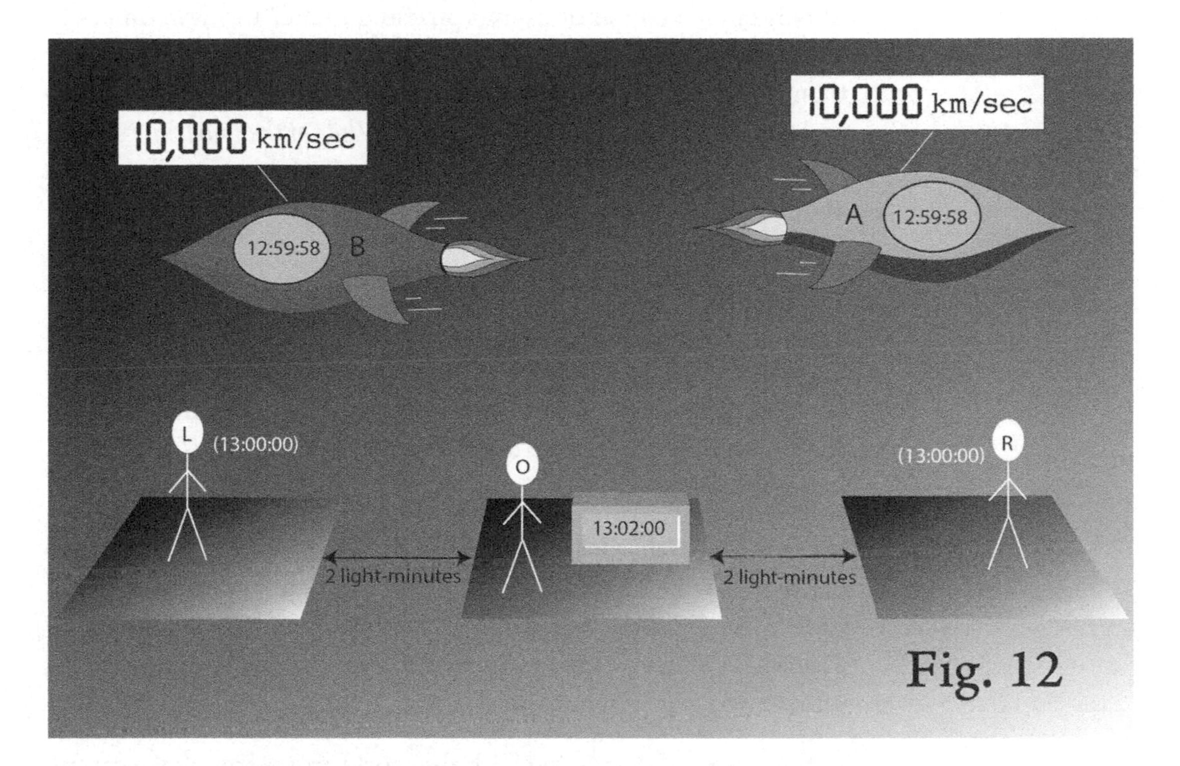

Fig. 12

A space ship carrying observer A whizzes past O at 10,000 km/sec (c/30) moving from left to right. At the same time a space ship carrying observer B whizzes past O at the same 10,000 km/sec, but moving from right to left.

Applying what we've learned about time dilation and the Lorentz factor, from O's perspective, clocks in *both* A's and B's spaceships must be running slow in the amount of two seconds per hour relative to O's clock.

However, from A's perspective, B's clock must be running slow by eight seconds per hour and from B's perspective, A's clock must be running slow by eight seconds per hour; that's because their speed, *relative to each other*, is nearly 20,000 km/sec (c/15). Furthermore, from both A's and B's perspectives, O's clock is running slow by two seconds per hour.[37] Those facts would appear to be contradictory since O concludes that A's and B's clocks are running at the same speed (i.e., each running two seconds per hour slower than O's), yet each of A and B concludes that the other's clock is running slow relative to their own!

This is often a contentious and puzzling matter for newcomers to relativity. Ask a physicist how it can be that A's clock is running slower than B's, while at the same time B's clock is running slower than A's, and the answer will be something like: "*While at the same time* is not a correct representation of the situation; the clocks are actually located at different space-time co-ordinates by virtue of their relative speed".

Got that?

Possibly not. And if you had nothing better to do with your time, that same physicist would be happy to prove to you that there is no contradiction or paradox here, but the proof would involve lots of mind-numbing space-time diagrams, which we've chosen to avoid in this book.

So, let's approach it from another angle. If you were A in the above discussion, why should you care what O infers about the rate at which time is passing in your spaceship? Or if you were B, why should you care what O infers? Perhaps you'd both be mildly interested, but it's unlikely to keep either of you awake at night.

In the course of interacting with the world around us, all we can really do is infer what's happening from our individual perspectives as observers and act on the result. Note *infer*, rather than see, hear, or measure. We may see lightning strike some distance away, and knowing

37. Mathematically dexterous readers can confirm these two-seconds-per-hour and eight-seconds-per-hour results by inserting 1/30 and 1/15 for k in the Lorentz equation, since 10,000 km/sec is 1/30 of the speed of light (c/30), and 20,000 km/sec is 1/15 the speed of light (c/15). At these relatively small fractions of the speed of light, every doubling of speed causes a four times increase in time dilation; as a mathematician would say, at speeds much less than c, the relationship between speed and time dilation is very close to a square law.

the speed of sound in air to be about a mile in five seconds, we begin counting off the seconds until we hear the crack or boom. If it takes five seconds for the sound to reach us, we can correctly infer that the strike occurred about a mile away from our location. People a different distance from the strike zone may reach a count of twenty seconds before hearing the sound, and if they also know the speed of sound in air, they would infer that the strike occurred about four miles away from where they are; and they also would be correct.

In our present discussion the fact in question is not the speed of sound in air, but the slowing of time for moving objects in accordance with the Lorentz factor, and how it can be that two observers in relative motion each see the other's clocks to be running slow.

The key point is that under the Principle of Relativity, each person's individual inferences are decisive if that person is trying to understand what's happening and make accurate predictions based upon it. It matters not a whit if others arrive at different conclusions based on their particular perspectives.

To get a better feel for this, let's place two more space platforms a distance of two light-minutes away[38] in the direction of travel of A and B, one off to O's left and the other off to O's right. (See again Fig. 12 on p. 99)

On each of these platforms is another observer, L on the left platform and R on the right. The three platforms are stationary relative to each other and their clocks are synchronized. By prior agreement, when A's space ship passes directly in front of R, A informs R by radio what time it is by his clock. R then sends a radio signal to O saying "When A passed by me, the time on his clock was such and such." Likewise, when B's spaceship passes L, B informs L by radio what the time is on B's clock. L then sends a radio signal to O saying "When B passed me, the time on his clock was such and such."

Let's say that at precisely 12:00:00 hours, O measures A and B passing his space platform traveling at 10,000 km/sec ($c/30$). O therefore can

38. Recall that a light-minute is the *distance* light travels in one minute; therefore, it would take a light flash two minutes to travel a distance of two light-minutes.

infer that by his own clock, it will take exactly one hour for A to reach R and for B to reach L (thirty times two minutes). But O knows about time dilation and the Lorentz factor, so he *infers* that time in both A's and B's spaceships is running slow by 2 seconds per hour relative to all three stationary platform clocks. So, he is not in the least surprised to receive a radio message from R shortly after 13:02:00, saying that the time on A's clock when it passed by just now was 12:59:58, and simultaneously a message from L saying that the time on B's clock when it passed by was also 12:59:58.[39] Not only is O not surprised, he could have predicted it and placed a winning bet on the outcome!

The foregoing is a long-winded way of saying that regardless of the seeming paradox that each of O, A and B will infer the others' clocks to be running slower than their own clock, if each makes predictions based on their own inferences, those predictions will always turn out to be correct; provided, of course, that they have correctly accounted for all factors that could affect the result.

What more could one reasonably ask of a scientific theory than that its predictions accurately reflect reality? The Special Theory of Relativity reflects reality to a very high degree indeed.

39. Readers might like to confirm those times for themselves, by noting that it takes an hour by O's clock for each of A and B to reach R and L, but two seconds less by A's and B's clocks, and that it takes 2 minutes for radio signals from R and L to reach O.

PROOF OF THE REALITY OF TIME DILATION

Now, back to young Albert in Bern in 1905. A prediction as unexpected and startling as time dilation for moving objects demanded experimental verification. However, because of the extremely high speeds needed to cause significant amounts of time dilation and the unavailability of clocks precise enough to measure it, it was many decades after the publication of Einstein's theory before proofs began to be discovered.

The first experimental confirmation of time dilation involved subatomic particles called mu-mesons, which are created in various nuclear reactions. They have a mean lifetime of 2.2 microseconds, meaning that on average, they will decay into other particles in that amount of time.[40]

Mu-mesons are also created in abundance in Earth's upper atmosphere by incoming cosmic rays (usually protons) smashing into the nuclei of oxygen atoms; these mu-mesons can readily be detected in ground based laboratories. Even though they travel very close to the speed of light, because of their short lifespans – and absent time-dilation effects – the average mu-meson would travel only 600 meters or so before decaying. In practice a much larger percentage survives the multi-kilometer journey from their point of creation in the upper atmosphere to the ground-based detectors. This is because their lifetimes are greatly extended due to relativistic time dilation caused by their high speeds. In the 1960s, a number of experiments were designed and carried out specifically to confirm the time dilation prediction of Special Relativity. Time dilation factors as high as fifty-to-one were recorded and were in excellent agreement with those predicted by the Lorentz formula based on the measured speed of incoming mu-mesons.

40. Some sub-atomic particles have mean lifetimes measured in picoseconds, whereas neutrons (found in the nucleus of every atom except the lightest isotope of hydrogen) have mean lifetimes of about 15 minutes in their free state and apparently last forever when bound up within atomic nuclei.

Later in the 20th century, as very precise atomic clocks were developed, it became possible to test time dilation predictions at the comparatively much lower speeds of commercial aircraft. One such experiment placed an atomic clock in an aircraft and flew it around the world for a day. Its speed was only about one-millionth the speed of light, at which speed the Lorentz formula predicted that its onboard Cesium-beam clock would lose about 43 nanoseconds per day relative to an identical ground-based clock. The experiment verified this to within the accuracy limits of the measuring apparatus.

In no small part because of its conflict with so-called common sense, during the latter part of the last century, hundreds of more and more sophisticated experiments were devised to confirm Einstein's prediction of time dilation, both the fact itself and its magnitude as calculated by the Lorentz factor; in all cases the prediction was confirmed to high accuracy.

Readers still skeptical about Einstein's and Liam's conclusions about time dilation are invited to come up with an alternative explanation that fits *all* the foregoing facts and also explains the results of the many experiments that have confirmed time dilation for moving objects. If you succeed, submit your explanation in the form of a peer-reviewed paper to *Annalen der Physik*, the same journal to which Einstein posted his original paper back in 1905. Then sit back and await a call from Stockholm in early October.

WHAT WE LEARNED IN SECTION III

- Via our most challenging thought experiment, we learned precisely why the Principle of Relativity in conjunction with the constancy of the speed of light imply that time in a moving frame passes at a slower rate relative to a frame deemed to be at rest.
- That the Lorentz equation allows one to calculate the amount of time dilation for a given relative speed.
- That there are real-world proofs of the reality of time dilation.

SECTION IV

IN WHICH WE CAREFULLY FOLLOW HARRY'S JOURNEY TO AND FROM CHANDRA CENTAURI, AND PROVE THAT HE'S THE ONE FOR WHOM TIME SLOWS DOWN, NOT CARRIE, AND THEREBY ALSO DEMONSTRATE THE FALLACY OF THE TWIN PARADOX.

MISSION TIME

We are now about to complete our stated purpose by applying what we've learned to follow Harry's mission to Chandra Centauri and back, and show conclusively that he would have aged only two years compared to Earth-based Carrie's ten years.

You will recall that in Harry's two-year (by his clock) space journey, his spaceship spent some time getting up to near light speed and getting back to stationary again, two times in all, in order to minimize the g-forces on the spaceship and its occupants. We arranged it like that to make his trip more credible, both scientifically and physiologically.

However, acceleration or deceleration of a moving object, in and of themselves, do not influence the passage of time (as inferred by a stationary observer) above and beyond the relativistic time dilation caused by the resulting speed. To clarify this, imagine that Lauren's spaceship in our last thought experiment starts from standstill beside the space platform and proceeds to accelerate at the enormous rate of 100,000 *g*. [Lauren wisely decides not to partake in this experiment; she's happy with her figure just as it is.] If we were to plot the speed of her spaceship over time, starting from standstill, after one second it would be traveling at about 1,000 km/sec, after two seconds it would be traveling at 2,000 km/sec, after three seconds it would be traveling at 3,000 km/sec, and so on. The instantaneous time dilation (i.e. the value at any particular moment as inferred by a stationary observer) would simply be equal to the value obtained by inserting the spaceship's speed *at that moment* into the Lorentz equation. And because the spaceship's speed is continually increasing, the time dilation effect would also increase moment by moment, exactly in accordance with the Lorentz equation, no more and no less. Suppose now that we arranged for the spaceship to come to a halt after one minute and we wish to know the accumulated time difference between a clock in the spaceship and a clock on the space platform. To do so, we'd have to invoke the mathematics of the calculus (integration), which isn't all that difficult, but it would add unnecessary complication to our effort to follow Harry's journey and keep track of the passage of time in his spaceship, since his journey involves four separate periods of acceleration and deceleration.

In order to avoid that complexity, we are going to make some small changes to Harry's journey, in particular, the way his spaceship accelerates and decelerates. In our love story, we talked about Harry's spaceship speeding up from stationary to 0.98c and slowing down as it approached Chandra Centauri, and the same on the return journey. We didn't specify how long it would spend speeding up and slowing down, because it's not relevant to what we're trying to accomplish. So, in the interests of numerical simplicity we'll specify that his space ship takes a negligible fraction of the total journey time speeding up and slowing down. Of course that means that Harry would be flattened thinner than a pancake by the enormous g-forces, but that just serves him right for abandoning his girl for ten (of her) years!

Here's the plan: We'll arrange for Harry's spaceship to immediately go from stationary to 98% the speed of light (.98c), then to execute an immediate turnaround when it reaches Chandra Centauri, returning to Earth at the same speed of .98c and finally to come to an immediate stop when it gets back to Earth. At .98c, the time dilation factor is almost exactly five to one, meaning that for every year that Harry ages, Carrie will age by five years, as required by our love story.

Readers will remember that the Principle of Relativity (which in effect says that what's good for the goose is good for the gander) implies that each should infer the other's clock to be running slower by virtue of their relative speed. So as Harry speeds off to Chandra Centauri at .98c, applying the Lorentz factor, Carrie should infer that time in Harry's spaceship is running at one fifth the rate of her Earth-time, and Harry should infer that time on Earth is running at one fifth the rate of his spaceship time. This is indeed what each would infer during the constant speed segments of Harry's journey.

Which brings us face to face with the dreaded Twin Paradox, which may be stated as follows:

> *Twin boys Mike and Spike decide to test the time travel predictions of special relativity. Mike heads out into space at .98c for a 10-year round trip journey, as measured by Spike's Earth-based clock. When*

> *Mike returns, he has aged by only two years, so Spike is now 8 years older than Mike. Why is it that the twin who sets out on the high-speed journey ages less than the stay-at-home twin? Doesn't the Principle of Relativity, by which each would observe time in the other twin's frame to be moving slower, put them on an equal footing?*

If you open a standard text on relativity and look for an explanation of the Twin Paradox, you are likely to learn that: "Only the traveling twin undergoes Inertial Frame Switching and that's why he ages less." Or: "The traveling twin must first accelerate in order to get up to speed and eventually decelerate in order to stop and compare clocks with the stay-at-home twin, and this breaks the symmetry of the situation." Both answers are correct, though not very helpful to readers unversed in the language of special relativity.

The fact is that the Twin Paradox is not a paradox at all; it's a misapplication of the Principle of Relativity, which readers will recall specifically applies only to uniform (non-accelerating) motion. Our Harry has to first accelerate to get up to .98c, then decelerate to stop and turn around, accelerate again to get up to .98c for the return journey and finally decelerate to a stop on arriving back on Earth. Harry is acutely aware of the forces acting on him that cause his spaceship to speed up and slow down, just as the passengers in a racecar would be. Carrie, on the other hand, remains on Earth and feels no such forces, going about her business as usual.

Our task therefore is to show that in spite of the Principle of Relativity (and the Twin Paradox), when Harry arrives back on Earth, he will have aged by only two years, while Earth-based Carrie will have aged by ten years (along with Harry's wine and his bonds).

We can do this quite easily by making use of the concepts in our last thought experiment in which we showed how the Principle of Relativity and the constancy of the speed of light causes time dilation.

Let's imagine that Harry's spaceship has the exact same clock/laser/porthole arrangement that Lauren's spaceship had and that there's a sheet of light-sensitive paper stretched all the way from Harry's point of departure near Earth to his turn-around point near Chandra Centauri (now *that's* a stretch). However, because Harry's spaceship will be traveling at .98c versus only .8c for Lauren's spaceship in our last thought experiment, the marks on the paper caused by the light pulses from the porthole will be 14.7 meters apart, versus 4 meters in Lauren's case.[41]

Thus, the laser in Harry's spaceship will leave marks 14.7 meters apart on the light-sensitive paper, all the way from Earth to Chandra Centauri. This could in principle be verified by having agents placed at various points along the way. However, since there will be no change in the speed of Harry's spaceship during his journey to Chandra Centauri, we really only need two agents; one located at Harry's departure point near Earth and one located at Harry's turn-around point near Chandra Centauri, 4.9 light years away. We provide both agents with perfectly accurate clocks that are synchronized to each other.[42]

Now, we're all set up to follow Harry's journey out and back:

At the stroke of midnight on December 31st, 2199, the crew of Harry's spaceship fires up the engines, instantly accelerates to .98c and heads for Chandra Centauri. The agent at the departure point confirms their departure speed in the same manner as Liam did in our earlier experiment. He also checks the light-sensitive paper near the departure point and notes that the marks are 14.7 meters apart, thus confirming that time for the spaceship and its occupants is now running at one fifth of Earth time.

41. Here's the explanation for that: At a speed of .98c an observer looking into Harry's spaceship would see a five-to-one slowing down of time, and that means that the light pulses would take 50 nanoseconds to get from the laser to the porthole. Traveling at .98c for 50 nanoseconds, the spaceship would therefore move 14.7 meters between light pulses. Readers might be surprised that by just increasing the spaceship's speed from .8c to .98c causes such a large increase in the amount of time dilation and in the spacing between the marks on the paper; that follows directly from the Lorentz equation, which we'll examine more closely in appendix 5.

42. By which we mean that if the readings of each clock were broadcast, an observer positioned exactly half-way between them would note that each clock indicated the same time. This is a valid way of confirming synchronization between clocks that are widely separated in space as long as there is no relative motion between them.

Five Earth-years later, just before midnight on December 31st, 2204, the agent at Chandra Centauri awaits the arrival of Harry's spaceship. He has set up apparatus to confirm the spaceship's incoming speed and its speed after turnaround. At the stroke of midnight, the spaceship arrives, does an instantaneous turn-around and departs back to Earth. The agent confirms its incoming and outgoing speeds to be .98c, then measures the spacing of the marks on the light-sensitive paper and notes that they are still 14.7 meters apart, both incoming and outgoing ones, as expected.

A few moments earlier, on board his spaceship, Harry's clock indicated the approach of midnight on December 31st, 2200. He saw Chandra Centauri up ahead, reached it and completed a turnaround at the stroke of midnight, then headed back to Earth at .98c.

After another five Earth years, just before midnight on December 31st, 2209, the agent at the Earth-orbiting station awaits the arrival of Harry's spaceship. He has set up apparatus to measure the spaceship's incoming speed. At the stroke of midnight, the spaceship arrives, right on time, and screeches to a halt. The agent confirms its incoming speed to be .98c, then measures the spacing of the marks on the light-sensitive paper and notes that they are 14.7 meters apart, as expected.

A few moments earlier, on board his spaceship, Harry's watch indicated the approach of midnight on December 31st, 2201. He saw the Earth-orbiting station up ahead and his spaceship arrived there at the stroke of midnight. He hit the brakes and screeched to a halt, mission completed.

Harry immediately checks his watch; it reads 00:00:00 on January 1st, 2202. He's been gone exactly two years by his watch and by his body's physiological clock. Then he checks the Earth-orbiting station's clock; it reads 00:00:00 on January 1st, 2210. He's been gone ten years of Earth time!

He calls his wine merchant to tell him that he'll be collecting his wine tomorrow. The merchant informs him that someone named Carrie picked up his wine yesterday. *Thoughtful of her; boy, am I ever looking forward to some of that fine California Cabernet.*

He calls his bank to instruct them to cash out his bonds. The teller informs him that someone named Carrie cashed them out yesterday. *Makes sense; their ten-year term was up a week ago. No point leaving the proceeds in the bank earning zero interest. Smart girl*!

Then he calls Carrie: No one picks up, her phone goes to auto-answer and Carrie's recorded voice invites a message. Harry complies; *Yo, Carrie. It's Harry. I'm back!*

Just as well it's a one-way connection. Otherwise Harry would have heard a slurred male voice in the background asking; *Harry? Who's Harry?*

We have now completed our task of proving that time dilation is a real physical phenomenon that is firmly grounded in scientific fact; that it would enable one-way travel to a future time if we had rocketry that could get us close enough to light speed, keep us there in relative comfort for a sufficient time, then slow us down so as to land safely at our chosen destination. Our destination could be back on Earth or some planet of a distant star.

Let's review our journey to this proof:

Our various thought experiments demonstrated that the speeds of bullets and sound waves are dependent on the speeds of their sources and/or the observers, but that light waves behaved differently. We learned of the empirical fact of the constancy of the speed of light for *all* observers, irrespective of the speed of the source of the light, and irrespective of observers' speed relative to the source of the light. We learned that this strange behavior of light could not be accounted for under Newton's laws of motion, which underpinned much of physics around the end of the 19th century. Specifically, since speed is defined as distance travelled in a given time, there was no way to account for the constancy of the speed of light without changing our understanding of both time and space. Scientists had to give up the notion of a constant time for all observers. This was Einstein's great epiphany.

We showed how Einstein arrived at this conclusion using our own thought experiment starring Lauren and Liam, in which Liam invoked the constancy of the speed of light in order to figure out what must be going on in Lauren's spaceship from his perspective. He concluded that time had to be passing more slowly in her spaceship relative to time on his space platform, and knowing the speed of her spaceship, he was able to calculate the amount of slowing down.

We added a twist to that thought experiment by which we allowed a skeptical Liam to confirm his calculations of what was actually going on, time-wise, in Lauren's spaceship. We did this by arranging for the spaceship's laser to make marks on a sheet of paper laid out just below the spaceship's travel path, and we related the spacing of those marks to the amount of time dilation occurring.

Applying these facts to Harry's journey, we concluded that an observer on Earth or on Chandra Centauri (which we presume to be stationary or moving relatively slowly with respect to Earth) would infer that time on Harry's spaceship was running at one-fifth of Earth time.

To confirm this, we arranged for agents at both ends of Harry's journey to examine the tracks left on the light-sensitive paper by the laser in Harry's spaceship. And, sure enough, the spacing between marks proved that time in Harry's spaceship was indeed running at one-fifth of Earth time *for the entirety of his journey*. We also explained that accelerations or decelerations of the traveler's spaceship in and of themselves have no effect on the passage of time on the spaceship as inferred by a stationary observer, so the fact that Harry's spaceship underwent massive acceleration and deceleration at the first and last moments of his outgoing and returning journeys did not affect this conclusion.

We could leave it at that, and justifiably claim that we have accomplished our stated goal.

But there are a few loose ends. How did Harry, being a physics teacher, reconcile the fact that he traveled a total distance of just under ten light years in only two years by both his own wristwatch and by his physiological clock? Sure, it took ten years by Earth clocks and by Carrie's physiological clock, but the Principle of Relativity gives Harry equal rights and obligations under the laws of physics, so to speak. And one of those laws states unequivocally that "*Thou cannot travel faster than the speed of light!*"

The problem is resolved when we consider the phenomenon of

Relativistic Length Contraction, which we alluded to briefly in one of our limericks and also in our love story.

Now we need to introduce it.

RELATIVISTIC LENGTH CONTRACTION

This (let's call it RLC), along with time dilation, the relativity of simultaneity, and the relationship between mass and energy, was one of the predictions of Special Relativity. RLC states that the length of an object in motion appears foreshortened *along its direction of motion* as seen or as measured by an observer deemed to be stationary. Equivalently, the distance between a traveler and his destination as measured by the traveler will be foreshortened according to the same principle. Moreover, the amount of foreshortening is given by the exact same Lorentz formula used to calculate time dilation. We just have to replace Time (t) with Length (l) in the formula.

To be clear about this, imagine that the object in motion is a perfect sphere when it's at rest relative to an observer. As it moves faster and faster across the observer's field of view it appears to get squeezed along its direction of motion *and only along its direction of motion*. As its speed approaches the speed of light, its original spherical shape will become more and more ellipsoidal, when viewed by an observer watching it pass by.

Note that this is not some kind of optical illusion; it is as real and as measurable as time dilation. Imagine that you are standing on a space platform somewhere between Earth and Chandra Centauri in order to observe Harry's spaceship pass by you from left to right, and let's say that his spaceship is 50 meters long according to both its manufacturer's specifications and also as measured by Harry from within the spaceship. If you could devise a way to measure the length of his spaceship as it passed you, you would measure it to be just 10 meters long; you would conclude that its length had reduced by the same factor (five) as time had slowed within the spaceship.

As we have learned, the rate of passage of time in a given frame, as measured by an observer at rest relative to that frame, is referred to as its Proper Time. Likewise, the length of an object in a given frame, as measured by an observer at rest relative to that frame, is referred to as its Proper Length.

To dispense with the thought that RLC might be some kind of illusion, imagine that the 50-meter long spaceship traveling at 0.98c has to pass through a tunnel that is wide and tall enough to accommodate its width and height, but is just a little more than 10 meters long. If you could take a photograph of the tunnel from the right perspective and at the right moment, the spaceship would be *completely* enveloped by the tunnel; none of it would be visible, all 50 meters of its (*proper*) length would be contained inside the 10-meter proper length of the tunnel!

The Special Theory has taught us that our assessments of the rate of passage of time, and of length along the direction of motion, are both speed dependent. Furthermore, the Principle of Relativity assigns as much validity to the views of a moving observer as one deemed to be at rest relative to what's being observed, since all motion is relative and there's no absolute state of rest.

Therefore, when an observer who's in motion relative to an object deemed to be at rest, infers the object to be of a certain length (which will always be less than its Proper Length), the observer's point of view is as valid as that of someone stationary beside the object in question. In particular, if the moving observer performs some calculation or other *based on his inferred length of the object* – as distinct from its Proper Length – *his* calculations will always give the correct answers in *his* frame, which he could, in principle, corroborate experimentally.

This RLC business may seem even stranger than time dilation. After all, time is an ephemeral thing, not something that you can grasp in your hand and examine. But the length of an object or the distance between two locations can be readily measured by a tape measure or a range finder in a manner that *everyone* would agree on: Yes?

No! Everyone who carries out the measurement *while stationary* with respect to the object or distance being measured would agree on the result. That's because they'd be measuring Proper Length. But if they carried out the measurement while in motion relative to what's being measured, they'd get a different answer according to the Lorentz formula.

Readers may be curious as to the cause of this length contraction. Back in 1889, the afore-mentioned Irish physicist George Francis

FitzGerald was trying to understand the null result of the Michelson-Morley experiment.[43] He was reluctant to give up on the idea of an all-pervasive ether, the (hypothesized) carrier of light waves, though the experiment's results strongly suggested that there was no ether. He theorized that an object moving through the ether at some speed would be contracted along its direction of motion by an ether wind,[44] and by just enough so as to make the ether undetectable by the experiment. Shortly thereafter, Hendrik Antoon Lorentz came to a similar conclusion and derived the exact formula that related the required amount of contraction to its speed; this became known as the Lorentz factor, while the effect itself is variously called the FitzGerald contraction, the Lorentz contraction, or the FitzGerald-Lorentz contraction, depending on the nationalistic biases of the writer; so the FitzGerald contraction it is!

When Einstein published his Special Theory of Relativity a few years later, length contraction in accordance with the Lorentz factor was one of the predictions that flowed directly from his two postulates; the ether played no part in it. Of course, not only would a spaceship be contracted in its direction of motion but all the people and contents therein would also be contracted in the same proportion, so far as a stationary observer is concerned. But the passengers wouldn't be in the least aware of it, just as they would be unaware of the slowing down of their time relative to an outside observer.

Does RLC seem a tad strange? It sure does, but it has the great charm that it comports perfectly with both theory and experiment.

Back to Harry. Because of RLC, he infers (and could measure with appropriate instrumentation) the distance between his spaceship and Chandra Centauri as being contracted by exactly the same factor as his time has become dilated, i.e. the Lorentz factor. And since speed is distance divided by time, he therefore sees no conflict between the reduced time needed for his journey according to his clock, the contracted distance to Chandra Centauri as measured by him, and

43. To review that experiment, see pp. 63-64.
44. It was assumed at the time that all objects in motion are moving through the ether, and by analogy with movement through the air, the objects would experience an *ether wind*.

his speed. It all computes perfectly.

RLC is a fascinating subject that has even more so-called paradoxes associated with it than does relativistic time dilation. Readers in a masochistic mood might care to google the subject or look it up on Wikipedia. But we're done with it here.

BUT, BUT...

Some readers may not be entirely convinced that we've fully explained how Harry was separated from Carrie for just two of his years whereas Carrie was separated from Harry for ten of her years. Sure, we showed quite definitely using our imaginary paper marks that Earth-based Carrie must conclude that time in Harry's spaceship ran at one fifth of Earth time for the entire duration of his journey. We also stated that the instantaneous reversal of direction of Harry's spaceship on reaching Chandra Centauri in and of itself had no effect on the passage of time for Harry from the perspective of an Earth-based observer. The unavoidable outcome is that Harry aged only two years to Carrie's ten years.

Fine you say, but what about Harry's perspective? He travelled at constant speed relative to Earth during every moment of his journey. We arranged for that in order to simplify our explanation. We did it by allotting zero time for his acceleration to 0.98c from Earth orbit, zero time for his turnaround at Chandra Centauri and zero time for coming to a stop on arriving back to Earth. Applying the Principle of Relativity, shouldn't Harry therefore infer that the clocks on Earth were running at one-fifth the rate of his spaceship's clocks during the entirety of his journey, just as Carrie had concluded that clocks in Harry's spaceship had been running at one-fifth the rate of her Earth clocks for the entirety of his journey? And if so, how do we square that with our conclusion that Harry aged only two years to Carrie's ten years? This, of course, is the central question posed by the Twin Paradox; namely how can it be that the traveling twin ages less than the stay-at-home twin since the Principle of Relativity insists that each twin can consider themselves to be at rest and that it's the other one who's the traveler?

One way to refute the Twin Paradox is to point out that the Principle of Relativity applies only to constant speed motion. And Harry had to undergo deceleration followed by acceleration in order to reverse direction and head back to Earth from Chandra Centauri. Therefore, his turn-around, no matter how much or how little time he took to execute it, makes the Principle of Relativity inapplicable.

Let's therefore ditch the Twin Paradox; it was useful to make us focus down on what's happening, and having done so, we see that there is no paradox.

So to recap, we've established that Carrie can prove that Harry's clock was running at one fifth the rate of her clock for the entirety of his journey, using our imaginary paper marks. But it seems reasonable for Harry to infer from the Principle of Relativity that Carrie's clock was running at one-fifth the rate of his clock for the entirety of his journey. But such an inference on his part can't be right since we've established that he ended up aging just two years to Carrie's ten years.

The answer must have something to do with Harry's turnaround, which is the one aspect of his journey that destroys the symmetry that would otherwise allow the Principle of Relativity to be applicable.

Actually it has everything to do with Harry's turnaround!

At a minimum, Harry will be subjected to (possibly enormous) g-forces as his rocket reverses direction, whereas Carrie feels no such forces, going about her business as usual on Earth. We stated earlier that Harry's deceleration and acceleration, in and of themselves do not affect Carrie's assessment of the rate at which time is passing in Harry's spaceship. But the reverse is not true. Someone undergoing acceleration or deceleration will themselves be subjected to time altering effects quite separate from any such effects resulting from their speed.

It is often asserted that since Special Relativity is restricted to situations involving constant speeds, a full appreciation of the consequences of Harry's turnaround requires a dive into the mathematics of General Relativity. But, thankfully, that's not so. It is certainly true that General Relativity can be used to fully explain why Harry aged less than Carrie. But without sweating the mathematics there's an easy way to show that his turnaround at Chandra Centauri introduces an asymmetry that justifies our assertion that the Principle of Relativity really does not apply here. And, additionally, it will be easy to see why that asymmetry causes Harry to agree with Carrie's assessment that on return to Earth, he will have aged less than Carrie.

Let's arrange for Harry and Carrie to keep track of the flow of time in each other's location for the duration of Harry's journey. We can do this (at least in principle) using television cameras, Carrie's camera pointed at a clock in Harry's spaceship and Harry's camera pointed at a clock in Carrie's house.

This is where our explanation of the Doppler effect in Section II has relevance. Readers will recall that the Doppler effect can give the illusion of time speeding up or time slowing down, depending on whether observers are approaching each other or receding from each other. Moreover, the Doppler effect is always larger than the effect of relativistic time dilation, at any given speed. This means that when observers are approaching each other, the apparent speeding up of time due to the Doppler effect is greater than the actual slowing down of time due to time dilation, so the net of the two is an *apparent* speeding up of time.

For the outgoing section of Harry's journey, which takes five years by Carrie's clock, they both see each other's clocks to be running slow, a combination of relativistic time dilation and the Doppler effect adding to each other. Moreover, Carrie will still see Harry's clock running slow for another 4.9 years by her clock *after* his turnaround at Chandra Centauri, because the first light of Harry's' approach takes a full 4.9 years to reach her. Carrie therefore sees Harry's clock to be running slow for 9.9 years of his 10-year journey and running fast only during the final 0.1 year.

For Harry, the outgoing journey takes one year by his clock. During all of that year he sees Carrie's' clock to be running slow, and for the same reasons that she sees his clock to be running slow., a combination of relativistic time dilation and the Doppler effect adding to each other. But on his turnaround he immediately sees Carrie's clock to be running fast since he's now approaching Earth. That's because the Doppler effect causes much more of an apparent speeding up than the time dilation effect causes a slowing down, and he will see this speeding up of Carrie's clock for the entire year (by his clock) of his return journey.

By the time Harry arrives home, Carrie will have observed fewer seconds elapse on his clock (because of the longer period during which it was running slow relative to running fast) than her own identical clock. So, Harry will have aged less than Carrie, according to Carrie. From Harry's perspective, the reverse is the case; he will have observed more seconds elapse on Carrie's clock because it will appear to be running faster than his own clock for the entirety of his return journey. So Carrie will have aged more than Harry, according to Harry. The net of it all is that Harry arrives home 10 years after leaving, according to Carrie's clock, having aged 8 years less than Carrie, and both would agree on that.

Note that in this section we did not set out to prove that Harry's observations will force him to conclude that he aged only two years to Carrie's ten years. Doing so would involve us in more mathematics than we care to impose on the reader. What we have shown is that Harry's perspective is definitely not the same as Carrie's because of his turnaround, and that as a consequence he will have aged less than Carrie.

Now we're done.

APPENDICES

1. MEASURING SPEED

Some of our thought experiments involved measuring the speed of a bullet or a sound wave or a light beam. Readers unfamiliar with physics or electronics might appreciate a basic understanding of how to make such measurements.

Let's say we want to measure the speed of a passing bullet. Consider the automatic doors found at store entrances: They often consist of an invisible (Infra-Red) narrow beam of light that crosses your path in front of the door and lands on a photocell (a device which generates an electrical current when illuminated). As you approach the door, your body interrupts the light beam and the electric current from the photocell drops to zero. An electronic circuit senses this change and causes the door to open.

We can use this principle to measure the speed of a bullet. Arrange two light beams each at right angles to the path of the bullet and, say, five meters apart (Fig. A1). Arrange a photocell to detect each beam as in the door-opener example. When the bullet crosses the path of the left beam, it will interrupt the light to the left photocell and cause its electrical current to drop to zero. Arrange for this change in current to *start* a precision electronic timer. When the bullet crosses the path of the right beam, arrange the drop in the right photocell's current to *stop* that timer. The readout of the timer thereby indicates how

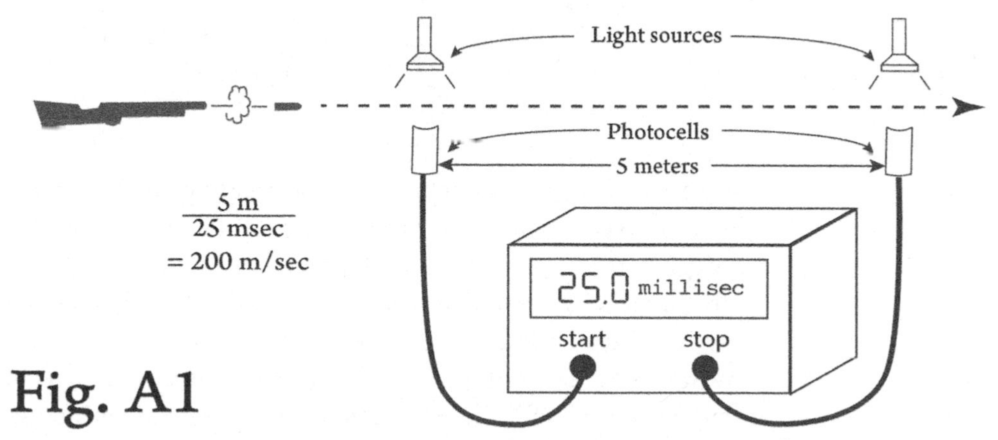

Fig. A1

long it took the bullet to traverse the distance between the two light beams. Divide that distance by the time indicated on the timer to get the bullet's speed. For example, if the distance between the beams is 5 meters and the timer reads 25 milliseconds (25 thousandths of a second), that would indicate a speed of 200 meters per second.

To measure the speed of a sound-wave, arrange two microphones a distance apart along the path of the sound wave (Fig. A2). When the front of the sound-wave hits the first microphone it will produce an electrical current (because that's what microphones do). Arrange for the onset of that current to *start* a precision timer. When the front of the sound-wave hits the second microphone arrange for the onset of its current to *stop* the timer. The readout of the timer thereby indicates how long it took the sound wave to travel the distance between the two microphones. Divide that distance by the time displayed on the timer to get the speed. For example, if the distance between the microphones is 5 meters and the timer reads 15 milliseconds, that would mean a speed of about 333 meters per second. (The nominal speed of sound in air at sea level is 340.29 meters per second)

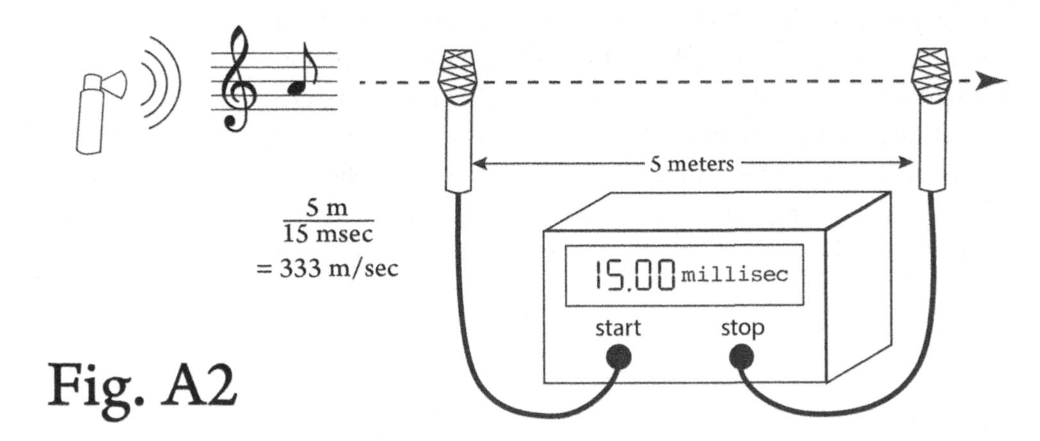

Fig. A2

To measure the speed of light, arrange a laser to generate a light flash (Fig. A3). Also arrange two mirrors some distance apart, each set at 45 degrees to the path of the light flash. The first mirror in the light's path is a partially silvered mirror, which reflects about half of the light downward to the left-side photocell and transmits the remainder onward to the second mirror, which reflects it downwards to the right-side photocell. When the light from the left-side mirror strikes

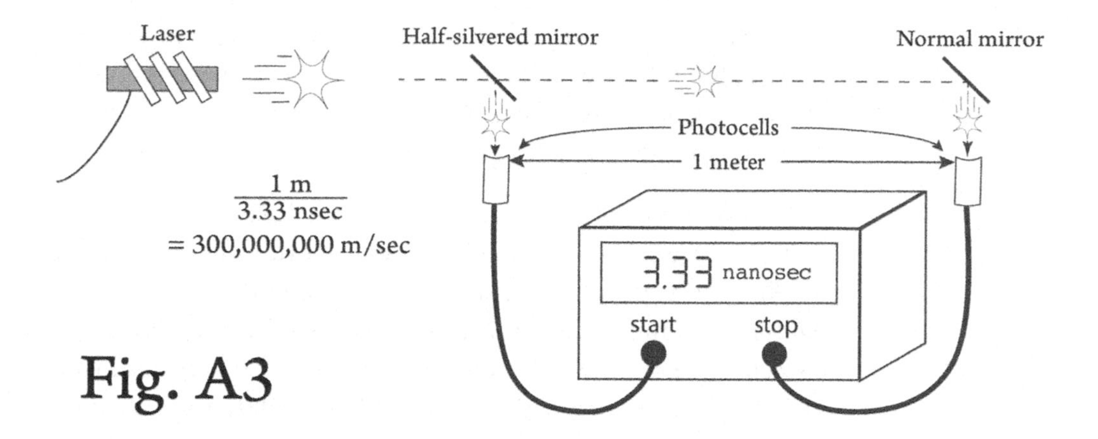

Fig. A3

the left photocell it generates an electric current in the photocell. Arrange for the onset of this current to *start* a precision electronic timer. When the light from the right-side mirror strikes the right photocell it also generates an electric current. Arrange for the onset of that current to *stop* the timer. The readout of the timer thereby indicates how long it took the light beam to traverse the distance between the two mirrors. Now divide that distance by the time displayed on the timer to get the speed of the light beam. For example, if the distance between the mirrors is 1 meter and the timer reads 3.33 nanoseconds (3.33 billionths of a second), that would indicate that the speed of the light flash is about 300,000,000 meters per second. Such a short time interval may seem way too small to measure, but modern electronic instruments are capable of parsing time and measuring the duration of events lasting mere trillionths of a second (picoseconds).

The speed of light in a vacuum is one of the fundamental and most important of physical constants, and techniques for measuring it have improved dramatically over the years. By 1975 its speed was known to be 299,792,458 meters per second, with a measurement uncertainty of just four parts in one billion, which means that its exact speed is somewhere between 299,792, 456.8 m/sec and 299,792,459.2 m/sec.

The meter was originally defined (in 1793) as one ten-millionth the distance from the equator to the North Pole. As a practical matter, this distance, nominally ten thousand kilometers, has a lot more uncertainty and variability than four parts in one billion (which amounts to a mere four centimeters). In later years, the meter was redefined as the length of a very precisely machined metal bar kept in a temperature and humidity-controlled chamber in Paris. This bar was used by the international scientific community as the ultimate reference for calibrating measuring devices worldwide.

Because the speed of light is such a fundamental feature of our universe and because we can now measure time with extraordinary precision, in 1983 it was decided to *define* and permanently set the speed of light in a vacuum at 299,792,458 meters per second *precisely*. As a consequence, the *meter* is now defined as the distance light travels in 1/299,792,458 of a second – (about 3.33564095 nanoseconds.

As measurement technologies improve over time, the speed of light will become even more precisely known, but since its nominal speed has been forever fixed at 299,792,458 meters per second, in the event that the extra precision is needed, *the length of the meter will be redefined to provide it*. So, some time in the future, the meter may be re-defined as (for example) the distance light travels in 1/299,792, 457.3 of a second, i.e. 3.33564096 nanoseconds.

2. COMPOSITION OF VELOCITIES: CO-LINEAR CASE

Here we prove that if a train is moving forward at a certain speed and if a bullet is fired forward from the train at a certain speed, then a stationary observer on the ground *outside the train* will observe (or measure) the bullet to be traveling at a speed equal to the sum of the train's speed and the speed with which the bullet left the rifle. Strictly speaking, the speed measured will be the *algebraic* sum of the two speeds, meaning that if the bullet were fired in the opposite direction to the train's speed, then the speeds would subtract. Furthermore, if the bullet were fired at some other angle relative to the direction of the train, then we'd have to dust off our trigonometry books to calculate the resultant speed as seen from the embankment. So, for simplicity we'll just assume that the bullet is fired in the same direction as the train is traveling.

Let's arrange for Lauren and her rifle to be inside a railway carriage that's 500 meters long. She's at the rear of the carriage, facing forward, and there's a target at the front end, 500 meters forward of her. Her rifle has a muzzle velocity of 500 km/hour.

The train is initially stationary. She fires a first bullet. Traveling at 500 km/hour the bullet, as we'd expect, will take 1/1000 of an hour (3.6 seconds) to reach the target 500 meters ahead of her within the carriage. Let the driver now start the train moving forward and when its speed reaches 100 km/hour let him keep it at that speed. Lauren now fires a second bullet. From our own experiences of traveling in trains we know that events happen inside a train in exactly the same manner whether the train is stationary or speeding straight along the tracks at constant speed. We could throw a ball back and forth to a fellow passenger, or we could juggle some objects or we could even play a game of table tennis, just as if the train were stationary. Also, we learned earlier that this fact is enshrined in the Principle of Relativity.

Therefore, from Lauren's perspective within the carriage, her second bullet will also take 3.6 seconds to reach the target inside the train. But the train is moving forward at 100 kilometers per hour relative to the embankment, so in the 3.6 seconds it takes the bullet to hit the bull's eye, the target, being attached to the carriage, has itself moved forward along the tracks a distance of 100 meters, so the total distance travelled by the bullet, from the perspective of an observer standing on the embankment, is therefore 600 meters. And if a bullet travels 600 meters in 3.6 seconds relative to an observer, its speed is therefore 600 km/hour relative to that observer. This speed, 600 km/hour, is the sum of the rifle's muzzle velocity (500 km/hour) and the train's speed (100 km/hour), which is what we set out to prove. Strictly speaking, as we'll see in appendix 6, the combined speeds would be the teeniest bit less than their simple sum, a direct consequence of special relativity.

3. COMPOSITION OF VELOCITIES: ORTHOGONAL CASE

This rule follows directly from the Principle of Relativity and from Newton's first law of motion, an empirical law which states that an object in motion will continue in motion with the same speed and in the same direction unless acted on by a force.

Let's return to the thought experiment with Lauren in her train carriage: Before the rifle is fired, the bullet in its chamber is stationary from Lauren's point of view within the carriage, but from Liam's point of view standing on the platform, it is moving horizontally at the same speed as Lauren's carriage.

On firing the rifle, the bullet is given a downward speed equal to the rifle's muzzle velocity. From Lauren's viewpoint within the carriage the bullet goes vertically downward and exits through the hole in the floor directly below the rifle's muzzle, exactly as one would expect. There are two things going on here:

First, the Principle of Relativity ensures that events in Lauren's speeding carriage will proceed in exactly the same manner as if her carriage were stationary with respect to the external platform. Thus, the bullet will always exit through the hole in the floor directly below the rifle's muzzle regardless of the carriage's speed (so long as that speed is constant).

Second, even though the bullet goes vertically downwards from Lauren's perspective, from Liam's platform perspective it still retains the carriage's horizontal speed along the railroad track, exactly as we'd expect from Newton's first law of motion. In other words, the explosive force that ejects the bullet from the rifle vertically downwards does not affect the *horizontal* speed that the rifle and bullet had by virtue of the train's speed.

So in order to plot the trajectory of the bullet from Liam's perspective, we calculate where the bullet will be at various times (t) after the rifle has fired. First calculate how far it will have moved horizontally by multiplying the train's speed by t, then calculate how far it will have moved vertically by multiplying the rifle's muzzle velocity by t. If you draw a horizontal line whose length is proportional to the train's speed and a vertical line whose length has the same proportion to the rifle's muzzle velocity, the diagonal will be the trajectory of the bullet from Liam's perspective.

4. INDEPENDENCE OF TIME DILATION FROM DIRECTION OF MOTION

In our earlier discussion of Newton's time versus Einstein's time, we asserted that the amount of relativistic time dilation caused by motion does *not* depend on whether the motion is toward, or away from, or at some other angle relative to the observer.

Here's justification for that: Let's return briefly to our last experiment and suppose that Liam has a toothache which needs immediate attention on the morning when Lauren in her spaceship is scheduled to make a high-speed pass by the platform location where his experiment is already set up. Liam's quarters are far away in another part of the vast sprawling space city and he has absolutely no way of getting there on time. So he calls his pal Mike who works at the space station and asks him to carry out the experiment in his place and to call him when he has calculated the amount of time dilation that Lauren must be experiencing as she speeds by at 80% of light speed. Mike carries out the experiment and duly phones Liam with the result. Since Liam's quarters are stationary relative to the space platform, there will be no difference in the flow of time between Liam and Mike. So, if Mike concludes that time in Lauren's spaceship is running at 60% of his platform time, it's also running at 60% of time in Liam's quarters, irrespective of the distance from and the angle at which Lauren's spaceship is passing Liam's quarters.

The point here is that regardless of the trajectory of a moving object relative to observer A, we can always place another observer B, *who is stationary relative to A,* in a position to observe the moving object passing directly in front of him and who would therefore arrive at the same conclusions arrived at by Liam in our time dilation thought experiment. And since A and B are stationary relative to each other, the flow of time will be identical for each of them. This proves that the magnitude of relativistic time dilation depends only on the speed of the moving object relative to the observer and is independent of both its direction of motion and its distance from said observer.

5. PROOF OF THE LORENTZ FORMULA

The Lorentz formula (or factor) defines the relationship between the rate at which time passes in a moving frame (e.g., a spaceship) relative to that of a stationary frame (e.g., a space platform) as a function of the speed of the moving frame. We can easily derive this factor with reference to our last thought experiment.

The vertical side of the triangle represents the distance between the laser and the porthole. This was 3 meters in that experiment and it took the light pulse 10 nanoseconds to travel that distance by Lauren's clock. Since we want to derive a general formula, let's represent that time by T_2.

Let k equal the speed of the spaceship relative to the speed of light. So, if the spaceship were traveling at 80% the speed of light (0.8c, where c represents the speed of light) then $k = 0.8$. Let T_1 be the time according to Liam's clock that the light pulse takes to get from the laser to the porthole.

Our task therefore is to derive a formula to allow us to calculate T_1 in terms of T_2 and k.

The length of the vertical side of the triangle is equal to cT_2, (distance is equal to speed multiplied by time). The horizontal side of the triangle represents the distance the spaceship traveled (as seen by Liam) during the flight of a light flash and its length is therefore equal to kcT_1 (again, distance is equal to speed (kc) multiplied by time T_1). The length of the diagonal, which is the path of the light flash, is simply cT_1.

Applying Pythagoras' theorem we get $(cT_2)^2 + (kcT_1)^2 = (cT_1)^2$ which simplifies to: $(T_1)^2 (1 - k^2) = (T_2)^2$. Thus $T_1 = T_2 / \sqrt{1 - k^2}$ which is the Lorentz formula. For our purposes, it's helpful to further rearrange it as: $T_2/T_1 = \sqrt{1 - k^2}$, in which T_2/T_1 is the factor by which time in Lauren's spaceship is running relative to time on Liam's space platform.

Returning briefly to our love story, if we put $k = 0.98$ which is Harry's speed, we see that the factor comes out to about 0.2, meaning that time for Harry is passing at one fifth the rate of that for Carrie, so in ten of Carrie's years, Harry will have aged by only two years.

Fig. A4 is a graph of $\sqrt{1 - k^2}$. The horizontal axis represents speed as a fraction of the speed of light, while the vertical axis represents the rate at which time is running in a spaceship traveling at that speed. Note that at 5% of light speed, which is 15,000 km/second (or 54,000,000 km/hour) – an enormous speed by present standards – the curve has barely begun to slope downwards, meaning that even at that speed the slowing down of time is negligible. The graph shows that time in a spacecraft traveling at 80% of light speed is running at 60% of normal, and at 98% of light speed, it shows that time is running at only 20% of normal, the number we used in our love story.

We've expanded the bottom right corner of the graph to show how dramatically time slows down as we travel closer and closer to the speed of light.

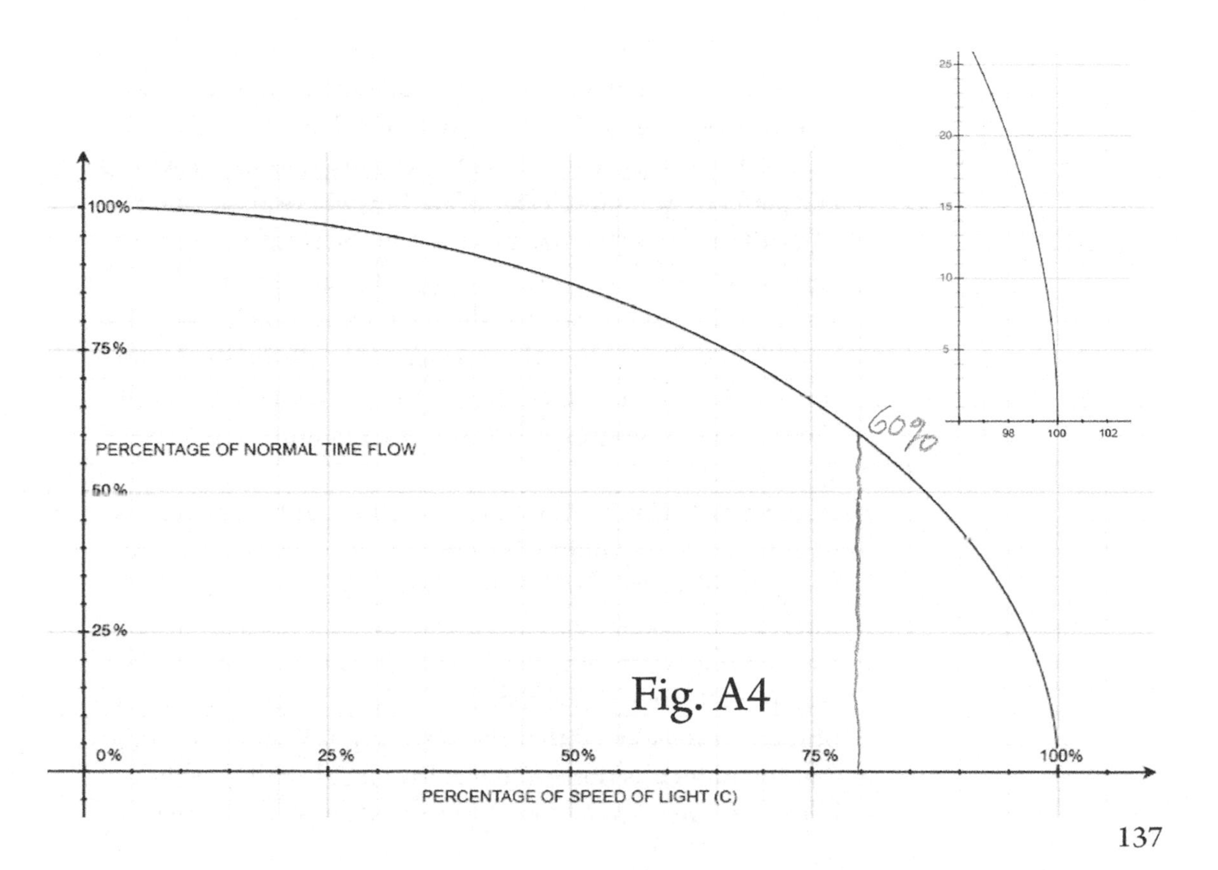

Fig. A4

6. NEWTON'S PHYSICS AND EINSTEIN'S PHYSICS

Though Isaac Newton died in 1726, the Newtonian worldview dominated physics until the early 20th century, when Einstein published first his special and later his general theories of relativity. It is an interesting historical fact that Einstein's special theory was a modification of Newton's laws of motion and his general theory was a modification of Newton's law of gravitation.

But just how much of a modification were they? There are two answers to this.

From the viewpoint of man's conception of the universe, the modifications were profound, startling and totally unanticipated. Here are some examples of how these world-views differed:

1. **In the Newtonian world-view, time was thought to be a constant and immutable stage on which the events of our lives played out and that its rate of passage was the same for everyone, irrespective of the city, planet, or galaxy they might live in. In other words, it was thought to be obvious that identical and perfectly accurate clocks located anywhere in the universe and initially synchronized to each other would proceed to tick away at the same rate and would ever after display the same time. Einstein's Special Theory showed conclusively that that is not the case; that the passage of time is relative rather than absolute and that it differs between clocks (and people, physical and biological processes, etc.) in motion relative to each other.**
2. **In the Newtonian world view, it was thought that the force of gravity was transmitted instantaneously between objects, however far apart they might be. Moreover, Newtonian physics offered no explanation of how the force of gravity could be transmitted across empty space. Einstein's General Theory interpreted gravity as a distortion of space-time by massive objects, and asserted that the resulting influence on other massive objects is transmitted at the large but finite speed of light. In fairness to Newton, he was well aware of the fact that**

he couldn't explain how gravity exerted itself across empty space over such long distances and that his assumption that gravity's influence was instantaneous was an unwarranted and possibly incorrect assumption. Nevertheless, his theory provided an extraordinarily accurate means to calculate the effects of gravity across the vast distances between planets, stars, and galaxies.

3. **In Newtonian physics one could, in principle, accelerate an object or vehicle to any desired speed up to and beyond the speed of light. Indeed, before Maxwell and Einstein, the speed of light didn't hold any more special significance for physicists than the speed of sound in air. In Einstein's physics the speed of light is a limiting speed in our universe. No object or signal can exceed it because, as the speed of an object approaches the speed of light, its momentum approaches infinity and so an infinite amount of energy would be required to accelerate it to the speed of light. Strictly speaking, special relativity does not forbid objects from traveling faster than the speed of light; it is more correct to say that special relativity forbids objects from being accelerated from lesser speeds up to the speed of light.**[45]

From the viewpoint of the application of these theories to science and engineering, the modifications were – and still are – largely miniscule or almost entirely irrelevant, except at extremes of speed (the speeds of subatomic particles) and extremes of mass (the masses of stars).

For example, in Newtonian physics, we saw in Lauren's and Liam's train experiments that the speed of a bullet fired forward from the front of a moving train would simply add to the speed of the train, from the point of view of an observer on the platform. This Newtonian *law of the addition* of velocities was assumed to hold true for any

45. So-called super-luminal particles (*which can only travel faster than the speed of light*) have been hypothesized by nuclear physicists, but no evidence nor reason for their existence has yet been found. On that score there is a "law" of physics proposed (somewhat tongue-in-cheek) by Murray Gell-Mann which asserts that *Everything That is Not Forbidden by Physical Laws is Compulsory*. We'll see – or our descendants will.

combination of the speeds of the train and the bullet. For example, if a train were traveling forward at speed X and if the rifle's muzzle velocity were Y, it was previously assumed that an observer on the platform would measure the bullet's speed as X+Y, even if that speed exceeded the speed of light.

In Einstein's physics, we must use a formula other than that of simple addition in order to determine the correct speed of the bullet as measured by a platform observer. The relativistic expression for this speed is: $(X+Y)/(1+XY/c^2)$. Readers up for a little mathematics will find that even if one sets both X and Y at the speed of light (c), the resultant speed would still be just c. They will also find that if X and Y are each significantly less than the speed of light, the formula gives almost exactly the same answer as the simple law of addition or subtraction that we used in our train experiments. For example, if X and Y are each 1,000,000 km/hour, a huge speed by current standards, the resultant speed as measured by the platform observer would be 1,999,978 km/hour, only about one part in 100,000 less than the speed predicted by just adding the two speeds together. And if the speeds of the train and the bullet were 1,000 km/hour, the error using simple addition would be less than one part in ten billion.

Here are some more specific examples:

1. **In Newtonian physics a clock in an orbiting spaceship traveling at about 28,000 km/hour would keep time exactly as a clock on the surface of Earth. In Einstein's physics the spaceship's clock would lose (only) about one second in one hundred years. (Time Dilation)**
2. **In Newtonian physics if the occupants of a spaceship speeding at 50,000 km/hour directly toward the moon 380,000 km away were to measure their distance from the moon, they would get the same result as if they were stationary. In Einstein's physics, they would measure the moon's distance as being closer by about the length of one's forearm. (Length Contraction)**
3. **In Newtonian physics, if one could measure the mass of that spaceship heading toward the moon, it would measure the same as it did when stationary on Earth. In Einstein's physics it would measure more by about 1 part in a billion. So if its**

mass were about the same as a fully-laden 747-400 aircraft (500,000 kilograms) it would measure a mere ½ gram more when moving at 50,000 km/hour ($M = E/c^2$).

The foregoing three examples come from Special Relativity. Here's one from General Relativity:

Up until the publication of Einstein's general theory in 1916, Newton's law of gravitation was able to predict and explain the speeds and orbits of all the planets with extraordinary accuracy – with a notable exception. The orbit of Mercury, the planet closest to the sun is more elliptical than that of the other planets. This ellipse rotates (precesses, more accurately) at a rate of about 1 degree every 8,000 years, a truly minuscule amount but still measurable with precision telescopes. That observation could not be explained by Newton's laws, but was explained convincingly by General Relativity. In fact, this was one of the early successes of the theory.

So, we can see that at the turn of the 20th century there was absolutely no need to question or improve on the accuracy of Newtonian physics, as its errors were far, far less than any instruments then available were capable of resolving. Special Relativity was developed because of the inability of Newtonian Physics to incorporate the constancy of the speed of light in its theoretical structures.

It is important to note that when a new scientific theory supplants an older one, as we're discussing here, the various facts and measurements that underpinned the older theory remain just as valid as they always were. New theories are needed when new facts are discovered that cannot be explained satisfactorily within the older theory. The constancy of the speed of light for all observers could not be accounted for in a theory in which time and space were independent and invariant entities. A new theory – Einstein's Special Theory of Relativity – was needed, and it encompasses all of the previously known facts along with the new fact of the constancy of the speed of light. And the price paid for that was the abandonment of the Newtonian conviction that time and space were independent and invariant, and their replacement by a combination of the two, called space-time.

Looking forward, it is the author's view that Special Relativity will survive the test of time (couldn't resist that!), but that's largely because at its heart, it's a simple theory and quite narrowly focused, i.e., restricted to uniform motion in gravity-free environments.

General Relativity, in contrast, was known to be incomplete within a decade or two of its conception when it became obvious that it had nothing to say at the atomic level where the force of gravity is overwhelmed by electromagnetic and nuclear forces. Hence the drive for a Theory of Everything, which is intended to seamlessly integrate General Relativity as applied to large-scale objects and Quantum Field Theory as applied at the subatomic level.

Nevertheless, General Relativity is one of the greatest intellectual achievements of humankind. Now over one hundred years old, it remains the go-to theory to explain everything from the large scale structure of the universe to the precession of the perihelion of Mercury, but it's as useless as *Finnegans Wake* at explaining the goings-on within atoms, notwithstanding the fact that Joyce's masterwork "predicted" quarks![46]

46. Murray Gell-Mann received the 1969 Nobel prize for showing that protons and neutrons are each composed of three smaller particles, which he called "quarks", from "Three quarks for Muster Mark", a line in *Finnegans Wake*.

7. THE GLOBAL POSITIONING SYSTEM (GPS)

The Global Positioning System (also known as Sat-Nav) deploys an array of 24 or more satellites in high Earth orbit, each with an orbital period of about 12 hours. Each satellite continuously transmits its current orbital position, *along with the time indicated by its onboard clock*. These transmissions are in microwave frequency bands which work well in all weather conditions.

This position and time information is used by mobile phones and computers to pinpoint location anywhere on Earth. GPS-enabled devices accomplish this by noting the arrival times of signals from at least four satellites and then performing calculations to compute distances from each satellite (trilateration[47]). From the results of these calculations, the location of GPS devices can be deduced with great precision. Receivers that utilize the more advanced features of the GPS can pinpoint location (including elevation) anywhere on Earth to an accuracy of a few meters.

The GPS satellites travel at a speed of 14,000 kilometers per hour at a height of 20,200 kilometers above Earth's surface. Because of their speed, relativistic time dilation would cause their on-board clocks to run about 3.5 microseconds per orbital period *slower* than ground based clocks.

There are actually two relativity-related effects that the GPS system has to allow for. The second (mentioned briefly earlier in this book) is caused by the slowing of time in a gravitational field relative to time in a lower or zero gravity environment, in accordance with Einstein's *general* theory of relativity. The stronger the gravity field the greater the time dilation. Because GPS satellites are orbiting at a height of 20,200 km above Earth's surface, their on-board clocks experience a weaker gravity field than ground-based clocks; this would cause the

47. Triangulation is the process of pinpointing a location by taking bearings from two or more locations. Trilateration is the process of pinpointing a location by measuring the distances to three or more points whose positions are known.

satellite-based clocks to run about 22.5 microseconds per orbital period *faster* than ground-based clocks.

The combination of these effects, if not corrected for, would cause each satellite's on-board clock to run about 19 microseconds per orbital period faster than ground-based clocks (22.5 - 3.5 = 19) when it's in orbit.

19 microseconds per twelve-hour orbit may seem negligible; that's less than one part in two billion or one second in seventy-two years. But bear in mind that microwaves, which travel at the same speed as light (300 meters per microsecond), will travel 5.7 kilometers in 19 microseconds. Since the arrival times of these microwaves at GPS receivers are used to compute location, the GPS system has to correct for this amount of clock error. If not corrected for, the clocks in GPS satellites placed in orbit would be ahead of ground based clocks by about 19 microseconds after just one orbit, causing your smartphones to misjudge your location by 5.7 kilometers! After two orbits the errors would be twice that. After a year or so, the errors would be so great that your GPS device couldn't tell whether you were in Timbuctoo, California or Timbuctoo, New Jersey.

The correction is carried out by adjusting the satellite-based clocks to run slow, *before launch*, by about 1.583 microseconds per hour (equal to 19 microseconds per 12-hour orbital period) so that *after* they're launched into orbit they'll run at the same rate as ground-based clocks.

The alternative would be to broadcast the correct ground time to the satellites at frequent intervals. How frequent? Well, since an error of 19 microseconds in time gives rise to an error of 5.7 kilometers in location accuracy, if we want an accuracy of say, 3 meters, we'd have to send corrections to the satellite clocks 1900 time per orbital period, i.e., about every 23 seconds! In practice, it's much, much easier to arrange for the on-board clocks to just run a teeny bit slower before launch.

Fantastically accurate as indeed they are, the on-board satellite clocks still need occasional corrections from ground-based control stations to maintain the long-term accuracy of the Global Positioning System.

It's amusing to speculate on what might have happened with the GPS system had Einstein not dreamed up his General Theory of Relativity before GPS was launched: It is widely believed that Einstein beat Lorentz and Poincaré by no more than a few months or years with his Special Theory of Relativity, as both these men were deep in the weeds of the problem and almost certainly one or other of them would also have solved it. But his General Theory was a whole different matter. That was very much the product of one man's mind in his quest for order in the universe and it took him a full ten years to complete the work. It wasn't a solution to a pressing problem as was the case with his Special Theory.

As we saw above, the timing error in the GPS system attributable to the satellites being high up in Earth's gravitational field was 22.5 microseconds per orbital period. The designers of the GPS system would not have been aware of that, absent General Relativity, so would only have built in a correction for time dilation due to the satellite's speed as specified by the Lorentz factor. On launch they would have discovered massive errors in location accuracy – nearly 7 kilometers per orbital period – with no idea of their cause. How long would it have taken them to figure out the cause of the errors and for the theoretical physics community to develop a theory to explain it? We'll never know.

8. TRAVELING BACKWARDS IN TIME

In a limited sense it is possible to travel back in time by deploying what we learned in the Introduction about gravity-induced time dilation. Here's how: Suppose the planet in which you and your ancestors live and have always lived is about the size of Earth and is in orbit around a super-massive black hole. Further, suppose that as a result of the intense local gravity caused by the black hole, time passes on this planet at one tenth the rate it does in gravity-free inter-galactic space.

If you had access to a sufficiently large and powerful space ship, you and a few thousand of your fellow-citizens could head out into space, well away from the black hole's gravity field and set up a new colony on some other convenient planet. Let's say you leave your ancestral planet on January 1st, 3000 and that you are 20 years old at the time and that your father, who decides to remain behind, is 50 years old.

Being free of the black hole's intense gravity field, time on your new planetary home will pass at ten times the rate it did (and continues to do) on the planet you left. Suppose after fifty years have passed in your new home – when you are 70 years old – nostalgia gets the better of you so you decide to travel back to your home planet. You leave on January 1st, 3050 local time, which would be January 1st, 3005 on your home planet. So, when you arrive "home" you will have effectively travelled backwards a full 45 years. If you meet up with your father he will be 55 years old to your 70 years; i.e., he would now be 15 years *younger* than you *in all physiological respects*. And assuming that scientific and technological research and development had progressed at much the same rate on both planets (as measured by their own clocks), you would be able to show him discoveries and technologies as yet unimagined in your ancestral home.

This form of traveling back in time is well within the realm of established science and involves none of the logical contradictions of Ms. Bright's journey. That's because such a time traveler would not be traveling back to his own past, so there is nothing he could do to alter the historical facts of the life he had already lived.

9. SCIENTIFIC THEORIES

A brief note on the use of the word 'theory' in scientific literature; in everyday usage it is taken to mean something unproven or speculative – a guess or hypothesis, perhaps. In scientific parlance, a Theory is an internally consistent body of knowledge that has acquired considerable experimental validity and is in accord with all known facts in its domain. Examples are the Theory of Quantum Electro-Dynamics, the Theory of Quantum Chromo-Dynamics, the Special Theory of Relativity, the General Theory of Relativity, and the Neo-Darwinian Theory of Evolution. The first two of these comprise the Standard Model of particle physics, which resulted from the collaborations of thousands of scientists over the course of six decades. The Higgs boson, predicted in 1964 and finally seen (statistically inferred, to be more precise) by the Large Hadron Collider in 2012, was a stunning example of the predictive power of the Standard Model.

The first four of these Theories (that capitalization is justified) form the bedrock of modern physics and underpin everything presently known about the physical universe, from the fundamental building blocks of atoms to the chemistry and biology of life and the large-scale structure of the known universe. But they are far from being the full story, as scientists are acutely aware; there is a lot of difficult theoretical work ahead before a Theory of Everything might be formulated, assuming such an all-encompassing theory is possible in the first instance – and that it is also within mankind's intellectual reach.

Readers may have heard about String theory, M-theory, Super-symmetry, and Loop Quantum Gravity. These are some of the names given to scientists' various and ongoing attempts at a Theory of Everything. But they are presently theories with a very small 't': far from complete, largely untested, but potentially containing the seeds of a final Theory of Everything.

Newton's Three Laws of Motion

The first law states that every object will remain at rest or continue traveling at constant speed in a straight line unless compelled to change its state of motion by the action of external forces. The key point is that if there is no net force acting on an object (if all the external forces cancel each other out) then the object will remain at rest or maintain a constant speed forever in whichever direction it was going in. If an external force is applied, the speed and direction of motion will change based on the magnitude and direction of the applied force.

The second law explains how the speed of an object changes when it is subjected to an external force. Specifically, if a constant force is applied to an object, the object undergoes a constant acceleration 'a' equal to the magnitude of the force 'F' divided by the mass 'm' of the object. In equation form, this is expressed as $a = F/m$, or alternatively as $F = ma$.

The third law states that when one body exerts a force on a second body, the second body simultaneously exerts a force equal in magnitude and opposite in direction on the first body. This law explains the production of thrust by a ship's propellers or by a jet or rocket engine.

Newton's Law of Universal Gravitation

This is Newton's greatest discovery, which he deduced mathematically from Kepler's observations of the motions of the planets.

The law states that every object in the universe exerts a force on every other object in the universe. The magnitude of that force depends on the masses of the objects and the distance between them. Thus, two objects of mass m_1 and m_2 a distance d apart will feel a force of attraction between them that's given by the relationship:

F is proportional to $(m_1 m_2)/d^2$

Thus, if you halve the distance between the objects, the strength of the attractive force will *increase* by factor of four and, conversely, if you double the distance between them the force will *decrease* by a factor of four. This is an example of an "inverse square" law, quite common in physics.

11. Q&A TIME

Q1. Did Harry Travel Faster than Light Speed on his Journey to Chandra Centauri?

In our love story, Harry travelled the 9.8 light-years distance to Chandra Centauri and back in just two years as measured by his wristwatch and also as experienced through his body's aging process.

So, why is he not entitled to assert that he travelled at nearly five times the speed of light?

Well, he is, but only by cherry-picking from these facts:

Fact 1. The distance from Earth to Chandra Centauri is 4.9 light-years. This is the proper distance, which you would measure if you slung a longish tape measure between the two places.

Fact 2. The distance from Earth to Chandra Centauri would be just 0.98 light-years according to Harry if he were able to measure that distance just after starting his journey when he'd be traveling at 98% of the speed of light.

Fact 3. The time it took Harry to travel out to Chandra Centauri and return back to Earth was 5 years according to Carrie, who stayed behind.

Fact 4. The time it took Harry to travel out to Chandra Centauri was 1 year, according to Harry's wristwatch and his on-board clocks.

Combining Facts 1 and 4 would allow Harry to conclude that he had travelled at nearly 5 times the speed of light. But combining Facts 2 and 3 would force him to conclude that he had travelled at a little less than 1/5 the speed of light! Slow-poke.

The appropriate way to decide the matter is for Harry to use his own measurement of the distance he has travelled in combination with his own measurement of the time it took him; this would be in full accord with the Principle of Relativity, which considers measurements of time and distance made by someone in uniform motion to be as valid (in their frames) as the same measurements made by stationary observers. Or alternatively, he could use Carrie's assessment of the distance he travelled (obtained from local star charts) in combination with Carrie's measurement of the time taken. Both of these methods yield the conclusion that Harry travelled at .98c throughout his journey.

In the event that some readers might wish to rebut this interpretation by pointing out that the distance from Earth to Chandra Centauri is in fact 4.9 light-years (because all the astronomy books say so), consider this: Suppose a spaceship crammed with white dwarf star afficionados has just travelled from a distant region of our galaxy and has no knowledge of the "local" geography. It's passing close by Earth on its way to Chandra Centauri, and it's traveling at .98c relative to Earth. The captain is curious as to how far Chandra Centauri is from Earth, but having no local star charts, she decides to time the journey. Knowing the spaceship's speed relative to Earth, she'll be able to calculate the distance once she knows her arrival time. Well, by her clock it will take exactly one year to get there, traveling at 0.98c, so she must logically conclude that Chandra Centauri is just .98 light-years from Earth, not the 4.9 light-years believed by the local yokels.

Q2. Does Time Dilation really affect biological processes like cell division and aging, or are its effects limited to non-living objects like clocks and subatomic particles?

As we learned earlier, books on relativity usually demonstrate how traveling at near light speed causes time to slow down by means of a thought experiment involving a single photon or light flash bouncing vertically straight up and down between two mirrors. This "light clock" can be said to "tick" once each time the photon or flash completes an up-down traverse. When the light clock is set in motion horizontally across a stationary observer's field of vision, it is clear that the length of the (now-diagonal) path taken by the light to get from one mirror to the other is longer than the straight up-down path, and since the speed of light is constant for all observers it follows that this light clock must tick more slowly than when it was stationary, according to the observer.

Most readers can accept this example of time dilation in action in the case of the light clock, but have a harder time accepting that clocks of any design, whether simple pendulum clocks or sophisticated mechanical or electronic clocks, will also tick more slowly in a moving frame relative to stationary observers and by exactly the same amount as the light clock.

What then of biological "clocks" like human or animal hearts? *Clocks, you say?* Why, yes; even though the human heart doesn't beat as regularly as even the cheapest mechanical clock, knowing one's resting heart rate – say it's 75 beats per minute – one could make a fair guess at the passage of one hour by counting your heart beats until you get to 4500. And a little calculation will show that after about 3 billion beats, your heart would be close to calling it quits. So whether or not you accept that a heart is a kind of clock, you will surely agree that a heartbeat is a time-dependent process.

Which leads naturally to the questions: is the rate of a heartbeat also influenced by relativistic time dilation? And if so, would people traveling at high speed age more slowly than their stay-at-home brethren? Yes, and yes.

For skeptical readers, we offer the following justification for asking them to take the leap of faith from the slowing down of moving light clocks to space-travelers aging at different rates dependent on their speed:

We start by admitting that while the reality of time dilation has been experimentally demonstrated beyond any doubt in the case of clocks and subatomic particles, it has not been so demonstrated in the case of biological processes. The reason for this is that the amount of time dilation that occurs at the maximum speed to which we can presently accelerate a living organism (about 50,000 km/hour) is very small relative to light speed. Even after a ten-year voyage at that speed, the returning organism would be younger than its stay-at-home brethren by only a fraction of a second. It is technically feasible for an on-board atomic clock to keep track of such small time differences in order to compare its elapsed time with an identical ground-based clock on return to Earth. But there is presently no conceivable experiment that could be performed on returning astronauts (or even a cluster of bacteria for that matter) to see if they had aged bodily by a fraction of a second less than they otherwise would have aged had they remained behind on Earth.[48]

Though we can't (yet) prove it empirically, we can nevertheless make a compelling case that time dilation would indeed affect all time-dependent phenomena – including the rate at which biological organisms develop and age – and that they would do so exactly in accordance with the Lorentz equation.

Once again we invoke the Principle of Relativity, which asserts that the laws of physics are the same in all frames moving at uniform speed. For example, any experiment one might carry out in a spaceship travelling at near-light speed relative to Earth should yield the exact same result as the same experiment carried out while the spaceship is cruising slowly between Earth and a nearby star, or is stationary on the ground.

So, if the Principle of Relativity is universally true, it follows logically that time dilation applies equally to biological processes as it does to mechanical and electronic devices like clocks. To illustrate this, imagine we perform an experiment on Earth that measures the time taken for the population of a strain of bacteria to double under controlled laboratory conditions and suppose we choose a strain that takes about one hour to double as measured by local clocks.

We then take the same bacterial strain and experimental apparatus (clocks included) from our Earth laboratory on a fast ride in a spaceship and perform the same doubling experiment again and again, at different *constant* spaceship speeds relative to Earth. We should expect, therefore, that every one of the experiments will give the same answer, namely, that the population of bacteria always doubles in about one hour, *as measured within the spaceship by the same clocks.*

48. It is well known that other aspects of space travel can interfere with living organisms, the most well-known example being the weakening of bones due to prolonged exposure to low or zero-gravity environments. But even if such relatively strong factors were removed, we would still be unable to detect aging differences of a fraction of a second, or even days for that matter.

That can't come as much of a surprise to readers: How could these identical experiments *not* yield the same answer when the only difference between one experiment and the next is the speed of the spaceship relative to Earth, coupled with the fact that the subject bacteria have no way of "knowing" what that speed is, since there is no sensation of speed within a spaceship or other vehicle traveling at constant speed.

Moreover, of the more than 100 billion other stars and planets in our galaxy, we can know as a matter of statistical certainty that many of them will be close to stationary relative to our spaceship, whatever speed it may be going at, while others will be receding or approaching at just about every imaginable speed up to the speed of light. So even if the bacteria had some way of 'knowing' the speed of their spaceship, why would it be their speed relative to Earth that they would know, rather than their speed relative to one of these myriad other planets or stars?

Of course they wouldn't know, nor care; as we've learned earlier, speed has no real meaning unless it's in relation to some other object. Therefore:

a) Since the time taken for the bacterial population to double is always one hour as measured by the on-board clocks irrespective of the spaceship's speed, and

b) On return to Earth, the clocks on the spaceship will show less elapsed time than Earth-based clocks (because of time dilation),

It follows that Earthlings must conclude that the one hour of Earth time it took for the bacterial population to double corresponded to a shorter period of spaceship time, at the end of which the bacterial population was only part of the way to doubling. Let's say that the spaceship moves at 0.98c for the duration of one of the experiments, at which speed the slowing down factor would be five to one. If Earthlings could observe what was happening in the spaceship during the one hour it takes for the bacterial population to double on Earth, they would see that only 12 minutes would have elapsed in the spaceship. They would need to watch for five hours by their Earth-based clocks for the spaceship's clock to gain the one hour needed for the bacterial population in the spaceship to double.

So, absent compelling evidence to the contrary, we can justifiably conclude that *all* time-dependent phenomena are subject to time dilation just as clocks and subatomic particles are.

Q3. Could future space travelers use the principles of police radar to keep track of their speed and location on a journey to a distant star or galaxy?

No. Even if space travelers could see through a telescope a large planet, a light-year away, say, as a practical matter they could not measure their speed relative to it using a radar system. Why? Because radar works by sending out a burst of radio pulses that travel at the speed of light and then bounce back from the object of interest. By measuring the *change* in round trip times between individual pulses it is possible to compute relative speed. In this example, the space travelers would have to wait two years to get their speed measurement and even then, they would only have learned their speed relative to the planet at that moment, a year earlier, when their burst of radar pulses reflected off its surface.

Readers may wonder why future space travelers would even contemplate a journey to a distant star system if they would be unable to measure their speed or deduce their location relative to Earth or to their destination when out in interstellar space. Well, long before the invention of the GPS, systems based on the principles of inertial navigation were deployed in aircraft to keep track of their speeds and locations. Inertial navigation systems use onboard accelerometers to accurately measure the aircraft's accelerations in three dimensions; forward-backward, left-right and up-down. By keeping track of how the aircraft's acceleration changed with time in each dimension, a computer could readily calculate its speed and location relative to its starting position. Similar techniques could be used to guide future spacecraft to distant stars.

SUGGESTIONS FOR FURTHER READING:

Einstein: His Life and Universe
Walter Isaacson.

The early chapters provide a fine account of the historical development of Special Relativity and Einstein's role in it.

Relativity Visualized
Lewis C. Epstein

An excellent introduction to the various aspects of special and general relativity for beginners. A classic.

Special Relativity: For the Enthusiastic Beginner
David Morin

Highly recommended for readers wishing to dig (a lot) deeper into special relativity in all its aspects, not just time dilation. This involves quite a step-up in technical depth and requires competence in high-school mathematics and physics. But plenty of limericks to ease the journey!

WORLD SCIENCE FESTIVAL

REH-LUH-TI-VI-STIK

METER-
GRAM-
LITER-

Made in the USA
Monee, IL
29 August 2021